Tucholsky Wagner Zola Scott Sydow Schlegel
Turgenev Wallace Fonatne Freud
Twain Walther von der Vogelweide Fouqué Friedrich II. von Preußen
Weber Freiligrath Frey
Fechner Fichte Weiße Rose von Fallersleben Kant Ernst Richthofen Frommel
Hölderlin
Engels Fielding Eichendorff Tacitus Dumas
Fehrs Faber Flaubert
Eliasberg Ebner Eschenbach
Maximilian I. von Habsburg Fock Eliot Zweig
Feuerbach Ewald Vergil
Goethe Elisabeth von Österreich London
Mendelssohn Balzac Shakespeare Dostojewski Ganghofer
Lichtenberg Rathenau Doyle Gjellerup
Trackl Stevenson Tolstoi Hambruch
Mommsen Thoma Lenz Hanrieder Droste-Hülshoff
Dach Verne von Arnim Hägele Hauff Humboldt
Reuter Rousseau Hagen Hauptmann Gautier
Karrillon Garschin Defoe Hebbel Baudelaire
Damaschke Descartes
Hegel Kussmaul Herder
Wolfram von Eschenbach Dickens Schopenhauer Rilke George
Darwin Melville Grimm Jerome
Bronner Bebel Proust
Campe Horváth Aristoteles Voltaire Federer
Bismarck Vigny Barlach Heine Herodot
Gengenbach
Storm Casanova Tersteegen Gilm Grillparzer Georgy
Chamberlain Lessing Langbein Gryphius
Brentano Lafontaine
Strachwitz Claudius Schiller Kralik Iffland Sokrates
Bellamy Schilling
Katharina II. von Rußland Gerstäcker Raabe Gibbon Tschechow
Löns Hesse Hoffmann Gogol Wilde Gleim Vulpius
Luther Heym Hofmannsthal Klee Hölty Morgenstern Goedicke
Roth Heyse Klopstock Kleist
Luxemburg Puschkin Homer Mörike
La Roche Horaz Musil
Machiavelli Kierkegaard Kraft Kraus
Navarra Aurel Musset Lamprecht Kind Moltke
Nestroy Marie de France Kirchhoff Hugo
Laotse Ipsen Liebknecht
Nietzsche Nansen Ringelnatz
Marx Lassalle Gorki Klett Leibniz
von Ossietzky May vom Stein Lawrence Irving
Petalozzi Platon Knigge
Pückler Michelangelo Kock Kafka
Sachs Poe Liebermann Korolenko
de Sade Praetorius Mistral Zetkin

The publishing house tredition has created the series **TREDITION CLASSICS**. It contains classical literature works from over two thousand years. Most of these titles have been out of print and off the bookstore shelves for decades.

The book series is intended to preserve the cultural legacy and to promote the timeless works of classical literature. As a reader of a **TREDITION CLASSICS** book, the reader supports the mission to save many of the amazing works of world literature from oblivion.

The symbol of **TREDITION CLASSICS** is Johannes Gutenberg (1400 – 1468), the inventor of movable type printing.

With the series, tredition intends to make thousands of international literature classics available in printed format again – worldwide.

All books are available at book retailers worldwide in paperback and in hardcover. For more information please visit: www.tredition.com

tredition was established in 2006 by Sandra Latusseck and Soenke Schulz. Based in Hamburg, Germany, tredition offers publishing solutions to authors and publishing houses, combined with worldwide distribution of printed and digital book content. tredition is uniquely positioned to enable authors and publishing houses to create books on their own terms and without conventional manufacturing risks.

For more information please visit: www.tredition.com

Getting Gold: a practical treatise for prospectors, miners and students

J. C. F. (Joseph Colin Frances) Johnson

Imprint

This book is part of the TREDITION CLASSICS series.

Author: J. C. F. (Joseph Colin Frances) Johnson
Cover design: toepferschumann, Berlin (Germany)

Publisher: tradition GmbH, Hamburg (Germany)
ISBN: 978-3-8495-0854-8

www.tredition.com
www.tredition.de

Copyright:
The content of this book is sourced from the public domain.

The intention of the TREDITION CLASSICS series is to make world literature in the public domain available in printed format. Literary enthusiasts and organizations worldwide have scanned and digitally edited the original texts. tradition has subsequently formatted and redesigned the content into a modern reading layout. Therefore, we cannot guarantee the exact reproduction of the original format of a particular historic edition. Please also note that no modifications have been made to the spelling, therefore it may differ from the orthography used today.

CONTENTS

PREFACE
GETTING GOLD

CHAPTER I
CHAPTER II
CHAPTER III
CHAPTER IV
CHAPTER V
CHAPTER VI
CHAPTER VII
CHAPTER VIII
CHAPTER IX
CHAPTER X
CHAPTER XI
CHAPTER XII

PREFACE

Some six years ago the author published a small book entitled "Practical Mining," designed specially for the use of those engaged in the always fascinating, though not as invariably profitable, pursuit of "Getting Gold." Of this ten thousand copies were sold, nearly all in Australasia, and the work is now out of print. The London *Mining Journal* of September 9th, 1891, said of it: "We have seldom seen a book in which so much interesting matter combined with useful information is given in so small a space."

The gold-mining industry has grown considerably since 1891, and it appeared to the writer that the present would be a propitious time to bring out a similar work, but with a considerably enlarged scope. What has been aimed at is to make "Getting Gold" a compendium, in specially concrete form, of useful information respecting the processes of winning from the soil and the after-treatment of gold and gold ores, including some original practical discoveries by the author. Practical information, original and selected, is given to mining company directors, mine managers, quartz mill operators, and prospectors. In "Rules of Thumb," chapters XI. and XII., will be found a large number of useful hints on subjects directly and indirectly connected with gold-mining.

The author's mining experience extends back thirty years and he therefore ventures to believe with some degree of confidence that the information, original or compiled, which the book contains, will be found both useful and profitable to those who are in any capacity interested in the gold-mining industry.

J. C. F. J.

LONDON, November, 1896.

GETTING GOLD

CHAPTER I

INTRODUCTORY

GOLD is a name to charm by. It is desired by all nations, and is the one metal the supply of which never exceeds the demand. Some one has aptly said, "Gold is the most potent substance on the surface of our planet." Tom Hood sings:

> Gold, gold, gold, gold! Bright and yellow, hard and cold;
> Molten, graven, hammered, rolled, Heavy to get, and light to hold; Stolen, borrowed, squandered, doled.

That this much appreciated metal is heavy to get is proved by the high value which has been placed on it from times remote to date, and that it is light to hold most of us know to our cost.

We read no farther than the second chapter in the Bible when we find mention of gold. There Moses speaks of "the land of Havilah, where there is gold"; and in Genesis, chapter xxiv., we read that Abraham's servant gave Rebekah an earring of half a shekel weight, say 5 dwt. 13 grs., and "two bracelets of ten shekels weight," or about 4 1/2 ozs. Then throughout the Scriptures, and, indeed, in all historic writings, we find frequent mention of the king of metals, and always it is spoken of as a commodity highly prized.

I have sometimes thought, however, that either we are mistaken in the weights used by the Hebrew nation in early days, or that the arithmetic of those times was not quite "according to Cocker." We read, I. Kings x. and xli., that Solomon in one year received no less than six hundred and three score and six talents of gold. If a talent of gold was, as has been assumed, 3000 shekels of 219 grains each, the value of the golden treasure accumulated in this one year by the

Hebrew king would have been 3,646,350 pounds sterling. Considering that the only means of "getting gold" in those days was a most primitive mode of washing it from river sands, or a still more difficult and laborious process of breaking the quartz from the lode without proper tools or explosives, and then slowly grinding it by hand labour between two stones, the amount mentioned is truly enormous.

Of this treasure the Queen of Sheba, who came to visit the Hebrew monarch, contributed a hundred and twenty talents, or, say, 600,000 pounds worth. Where the Land of Ophir, whence this golden lady came, was really situated has evoked much controversy, but there is now a general opinion that Ophir was on the east coast of Africa, somewhere near Delagoa Bay, in the neighbourhood of the Limpopo and Sabia rivers. It should be mentioned that the name of the "black but comely" queen was Sabia, which may or may not be a coincidence, but it is certainly true that the rivers of this district have produced gold from prehistoric times till now.

The discovery of remarkable ruins in the newly acquired province of Mashonaland, which evince a high state of civilisation in the builders, may throw some light on this interesting subject.

The principal value of gold is as a medium of exchange, and its high appreciation is due, first, to the fact that it is in almost universal request; and, secondly, to its comparative scarcity; yet, oddly enough, with the exception of that humble but serviceable metal iron, gold is the most widely distributed metal known. Few, if any, countries do not possess it, and in most parts of the world, civilised and uncivilised, it is mined for and brought to market. The torrid, temperate, and frigid zones are almost equally auriferous. Siberia, mid-Asia, most parts of Europe, down to equatorial and southern Africa in the Old World, and north, central, and southern America, with Australasia, in what may be termed the New World, are all producers of gold in payable quantities.

In the earlier ages, the principal source of the precious metal was probably Africa, which has always been prolific in gold. To this day there are to be seen in the southern provinces of Egypt excavations and the remains of old mine buildings and appliances left by the ancient gold-miners, who were mostly State prisoners. Some of these mines were worked by the Pharaohs of, and before, the time of Moses; and in these dreadful places thousands of Israelites were driven to death by the taskmaster's whip. Amongst the old appliances is one which approximated very closely to the amalgamating, or blanket table, of a modern quartz mill.

The grinding was done between two stones, and possibly by means of such primitive mechanism as is used to-day by the natives of Korea.

The Korean Mill is simply a large hard stone to which a rocking motion is given by manual power by means of the bamboo handles while the ore is crushed between the upper and basement stone.

Solomon says "there is no new thing under the sun"; certainly there is much that is not absolutely new in appliances for gold extraction. I lately learned that the principle of one of our newest concentrating machines, the Frue vanner, was known in India and the East centuries ago; and we have it on good authority—that of Pliny—that gold saving by amalgamation with mercury was practised before the Christian era. It will not be surprising then if, ere long, some one claims to have invented the Korean Mill, with improvements.

Few subjects in mineralogical science have evoked more controversy than the origin of gold. In the Middle Ages, and, indeed, down to the time of that great philosopher, Sir Isaac Newton, who was himself bitten with the craze, it was widely believed that, by what was known as transmutation, the baser metals might be changed to gold; and much time and trouble were expended in attempts to make gold—needless to say without the desired result. Doubtless, however, many valuable additions to chemical science, and also some useful metallic alloys, were thus discovered.

The latest startling statement on this subject comes from, of course, the wonderland of the world, America. In a recently published journal it is said that a scientific metallurgist there has succeeded in producing absolutely pure gold, which stands all tests, from silver. Needless to say, if this were true, at all events the much vexed bi-metallic question would be solved at once and for all time.

It is now admitted by all specialists that the royal metal, though differing in material respects in its mode of occurrence from its useful but more plebeian brethren of the mineral kingdom, has yet been deposited under similar conditions from mineral salts held in solution.

The first mode of obtaining this much desired metal was doubtless by washing the sand of rivers which flowed through auriferous strata. Some of these, such as the Lydian stream, Pactolus, were supposed to renew their golden stores miraculously each year. What really happened was that the winter floods detached portions of auriferous drift from the banks, which, being disintegrated by the rush and flow of the water, would naturally deposit in the still reaches and eddies any gold that might be contained therein.

The mode of washing was exactly that carried on by the natives in some districts of Africa to-day. A wooden bowl was partly filled with auriferous sand and mud, and, standing knee-deep in the stream, the operator added a little water, and caused the contents of the bowl to take a circular motion, somewhat as the modern digger does with his tin dish, with this difference, that his ancient prototype allowed the water and lighter particles to escape over the rim as he swirled the stuff round and round. I presume, in finishing the operation, he collected the golden grains by gently lapping the water over the reduced material, much as we do now.

I have already spoken of the mode in which auriferous lode-stuff was treated in early times—i.e., by grinding between stones. This is also practised in Africa to-day, and we have seen that the Koreans, with Mongolian acuteness, have gone a step farther, and pulverise the quartz by rocking one stone on another. In South America the arrastra is still used, which is simply the application of horse or mule power to the stone-grinding process, with use of mercury.

The principal sources of the gold supply of the modern world have been, first, South America, Transylvania in Europe, Siberia in Asia, California in North America, and Australia. Africa has always produced gold from time immemorial.

The later development in the Johannesburg district, Transvaal, which has absorbed during the last few years so many millions of English capital, is now, after much difficulty and disappointment—

thanks to British pluck and skill—producing splendidly. The yield for 1896 was 2,281,874 ounces—a yield never before equalled by lode-mining from one field.

In the year 1847 gold was discovered in California, at Sutor's sawmill, Sacramento Valley, where, on the water being cut off, yellow specks and small nuggets were found in the tail race. The enormous "rush" which followed is a matter of history and the subject of many romances, though the truth has, in this instance, been stranger than fiction.

The yield of the precious metal in California since that date up to 1888 amounts to 256,000,000 pounds.

Following close on the American discovery came that of Australia, the credit of which has usually been accorded to Hargraves, a returned Californian digger, who washed out payable gold at Lewis Ponds Creek, near Bathurst, in 1851. But there is now no reason to doubt that gold had previously been discovered in several parts of that great island continent. It may be news to many that the first gold mine worked in Australia was opened about twelve miles from Adelaide city, S.A., in the year 1848. This mine was called the Victoria; several of the Company's scrip are preserved in the Public Library; but some two years previous to this a man named Edward Proven had found gold in the same neighbourhood.

Most Governments nowadays encourage in every possible way the discovery of gold-fields, and rewards ranging from hundreds to thousands of pounds are given to successful prospectors of new auriferous districts. The reward the New South Wales authorities meted out to a wretched convict, who early in this century had dared to find gold, was a hundred lashes vigorously laid on to his already excoriated back. The man then very naturally admitted that the alleged discovery was a fraud, and that the nugget produced was a melted down brass candlestick. One would have imagined that even in those unenlightened days it would not have been difficult to have found a scientist sufficiently well informed to put a little nitric acid on the supposed nugget, and so determine whether it was the genuine article, without skinning a live man first to ascertain. My belief is that the unfortunate fellow really found gold, but, as Mr. Deas Thompson, the then Colonial Secretary, afterwards told

Hargraves in discouraging his reported discovery, "You must remember that as soon as Australia becomes known as a gold-producing country it is utterly spoiled as a receptacle for convicts."

This, then, was the secret of the unwillingness of the authorities to encourage the search for gold, and it is after all due to the fact that the search was ultimately successful beyond all precedent, that Australia has been for so many years relieved of the curse of convictism, and has ceased once and for all to be a depot for the scoundrelism of Britain—"Hurrah for the bright red gold!"

Since the year 1851 to date the value of the gold raised in the Australasian colonies has realised the enormous amount of nearly 550,000,000 pounds. One cannot help wondering where it all goes.

Mulhall gives the existing money of the world at 2437 million pounds, of which 846 millions are paper, 801 millions silver, and 790 millions gold. From 1830 to 1880 the world consumed by melting down plate, etc., 4230 tons of silver more than it mined. From 1800 to 1870 the value of gold was about 15 1/2 times that of silver. From 1870 to 1880 it was 167 times the value of silver and now exceeds it over twenty times. In 1700 the world had 301 million pounds of money; in 1800, 568 million pounds; and in 1860, 1180 million pounds sterling.

The gold first worked for in Australia, as in other places, was of course alluvial, by which is usually understood loose gold in nuggets, specks, and dust, lying in drifts which were once the beds of long extinct streams and rivers, or possibly the moraines of glaciers, as in New Zealand.

Further on the differences will be mentioned between "alluvial" and "reef" or lode gold, for that there is a difference in origin in many occurrences, is, I think, provable. I hold, and hold strongly, that true alluvial gold is not always derived from the disintegration of lodes or reefs. For instance, the "Welcome Nugget" certainly never came from a reef. No such mass of gold, or anything approaching it, has ever yet been taken from a quartz matrix. It was found at Bakery Hill, Ballarat, in 1858, weight 2195 ozs., and sold for 10,500 pounds. This was above its actual value.

The "Welcome Stranger," a still larger mass of gold, was found amongst the roots of a tree at Dunolly, Victoria, in 1869, by two starved out "fossickers" named Deeson and Oates. The weight of this, the largest authenticated nugget ever found was 2268 1/2 ozs., and it was sold for 10,000 pounds, but it was rendered useless as a specimen by the finders, who spent a night burning it to remove the adhering quartz.

But the ordinary digger neither hopes nor expects to unearth such treasures as these. He is content to gather together by means of puddling machine, cradle, long tom, or even puddling tub and tin dish, the scales, specks, dust, and occasional small nuggets ordinarily met with in alluvial "washes."

Having sunk to the "wash," or "drift," the digger, by means of one or more of the appliances mentioned above, proceeds to separate the gold from the clay and gravel in which it is found. Of course in large alluvial claims, where capital is employed, such appliances are superseded by steam puddles, buddles, and other machinery, and sometimes mercury is used to amalgamate the gold when very fine. Hydraulicing is the cheapest form of alluvial mining, but can only be profitably carried out where extensive drifts, which can be worked as quarry faces, and unlimited water exist in the same neighbourhood. When such conditions obtain a few grains of gold to the yard or ton will pay handsomely.

Lode or reef mining, is a more expensive and complicated process, requiring much skill and capital. First, let me explain what a lode really is. The American term is "ledge," and it is not inappropriate or inexpressive. Imagine then a ledge, or kerbstone, continuing to unknown depths in the earth at any angle varying from perpendicular to nearly horizontal. This kerbstone is totally distinct from the rocks which enclose it; those on one side may be slate, on the other, sandstone; but the lode, separated usually by a small band of soft material known to miners as "casing," or "fluccan," preserves always an independent existence, and in many instances is practically bottomless so far as human exploration is concerned.

There are, however, reefs or lodes which are not persistent in depth. Sometimes the lode formation is found only in the upper and newer strata, and cuts out when, say, the basic rocks (such as granite, etc.) are reached. Again, there is a form of lode known among miners as a "gash" vein. It is sometimes met with in the older crystalline slates, particularly when the lode runs conformably with the cleavage of the rock.

Much ignorance is displayed on the subject of lode formation and the deposition of metals therein, even by mining men of long experience. Many still insist that lodes, particularly those containing gold, are of igneous origin, and point to the black and brown ferromanganic outcrops in confirmation. It must be admitted that often the upper portions of a lode present a strong appearance of fire agency, but exactly the same appearance can be caused by oxidation of iron and manganese in water.

It may now be accepted as a proven fact that no true lode has been formed, or its metals deposited except by aqueous action. That is to say, the bulk of the lode and all its metalliferous contents were once held in solution in subterranean waters, which were ejected by geysers or simply filtered into fissures formed either by the shrinkage of the earth's crust in process of cooling or by volcanic force.

It is not contended that the effect of the internal fires had no influence on the formation of metalliferous veins, indeed, it is certain that they had, but the action was what is termed hydrothermal (hot water); and such action we may see in progress to-day in New Zealand, where hot springs stream or spout above the surface, when the silica and lime impregnated water, reduced in heat and released from pressure, begins forthwith to deposit the minerals previously held in solution. Hence the formation of the wondrous Pink and White Terrace, destroyed by volcanic action some eight years since, which grew almost while you watched; so rapidly was the silica deposited that a dead beetle or ti-tree twig left in the translucent blue water for a few days became completely coated and petrified.

Gold differs in its mode of occurrence from other metals in many respects; but there is no doubt that it was once held in aqueous solution and deposited in its metallic form by electro-chemical action. It is true we do not find oxides, carbonates, or bromides of gold in Nature, nor can we feel quite sure that gold now exists naturally as a sulphide, chloride, or silicate, though the presumption is strongly that it does. If so, the deposition of the gold may be ceaselessly progressing.

Generally reef gold is finer as to size of the particles, and, as a rule, inferior in quality to alluvial. Thus, in addition to the extra labor entailed in breaking into one of the hardest of rocks, quartz, the *madre de oro* ("mother of gold") of the Spaniards, there is the additional labour required to pulverise the rock so as to set free the tiniest particles of the noble metal it so jealously guards. There is also the additional difficult operation of saving and gathering together these small specks, and so producing the massive cakes and bars of gold in their marketable state.

Having found payable gold in quartz on the surface, the would-be miner has next to ascertain two things. First, the strike or course of the lode; and secondly, its underlie, or dip. The strike, or course, is the direction which the lode takes lengthwise.

In Australia the term "underlie" is used to designate the angle from the perpendicular at which the lode lies in its enclosing rocks, and by "dip" the angle at which it dips or inclines lengthwise on its course. Thus, at one point the cap of a lode may appear on the surface, and some distance further the cap may be hundreds of feet below. Usually a shaft is sunk in the reef to prove the underlie, and a level, or levels, driven on the course to ascertain its direction underground, also if the gold extends, and if so, how far. This being proved, next a vertical shaft is sunk on the hanging or upper wall side, and the reef is either tapped thereby, or a cross-cut driven to intersect it.

We will now assume that our miners have found their lode payable, and have some hundreds of tons of good gold-bearing stone in sight or at the surface. They must next provide a reducing plant. Of means for crushing or triturating quartz there is no lack, and every year gives us fresh inventions for the purpose, each one better than that which preceded it, according to its inventor. Most practical men, however, prefer to continue the use of the stamper battery, which is virtually a pestle and mortar on a large scale. Why we adhere to this form of pulverising machine is that, though somewhat wasteful of power, it is easily understood, its wearing parts are cheaply and expeditiously replaced, and it is so strong that even the most perversely stupid workman cannot easily break it or put it out of order.

The stone, being pounded into sand of such degree of fineness as the gold requires, passes through a perforated iron plate called a "grating," or "screen," on to an inclined surface of copper plates faced with mercury, having small troughs, or "riffles," containing mercury, placed at certain distances apart.

The crushed quartz is carried over these copper "tables," as they are termed, thence over the blanket tables—that is, inclined planes covered with coarse serge, blankets, or other flocculent material—so that the heavy particles may be caught in the hairs, or is passed over vanners or concentrating machines. The resulting "concentrates" are washed off from time to time and reserved for secondary treatment.

To begin with, they are roasted to get rid of the sulphur, arsenic, etc., which would interfere with the amalgamation or lixiviation, and then either ground to impalpable fineness in one of the many triturating pans with mercury, or treated by chlorine or potassium cyanide.

If, however, we are merely amalgamating, then at stated periods the battery and pans are cleaned out, the amalgam rubbed or scraped from the copper plates and raised from the troughs and riffles. It is then squeezed through chamois leather, or good calico will do as well, and retorted in a large iron retort, the nozzle of which is kept in water so as to convert the mercury vapour again to the metallic form. The result is a spongy cake of gold, which is either sold as "retorted" gold or smelted into bars.

The other and more scientific methods of extracting the precious metal from its matrices, such as lixiviation or leaching, by means of solvents (chlorine, cyanogen, hyposulphite of soda, etc.), will be more fully described later on.

CHAPTER II

GOLD PROSPECTING — ALLUVIAL AND GENERAL

It is purposed in this chapter to deal specially with the operation of searching for valuable mineral by individuals or small working parties.

It is well known that much disappointment and loss accrue through lack of knowledge by prospectors, who with all their enterprise and energy are often very ignorant, not only of the probable locality, mode of occurrence, and widely differing appearance of the various valuable minerals, but also of the best means of locating and testing the ores when found. It is for the information of such as these that this chapter is mainly intended, not for scientists or miners of large experience.

All of us who have had much to do with mining know that the majority of the best mineral finds have been made by the purest accident; often by men who had no mining knowledge whatever; and that many valuable discoveries have been delayed, or, when made, abandoned as not payable, from the same cause—ignorance of the rudiments of mineralogy and mining. I have frequently been asked by prospectors, when inspecting new mineral fields, what rudimentary knowledge will be most useful to them and how it can be best obtained.

If a man can spare the time a course of lessons at some accredited school of mines will be, undoubtedly, the best possible training; but if he asks what books he should read in order to obtain some primary technical instruction, I reply: First, an introductory text-book of geology, which will tell him in the simplest and plainest language all he absolutely requires to know on this important subject. Every prospector should understand elementary geology so far as general knowledge of the history of the structure of the earth's crust and of the several actions that have taken place in the past, or are now in operation, modifying its conditions. He may with advantage go a few steps further and learn to classify the various formations into systems, groups, and series: but he can acquire all that he need absolutely know from this useful little 2s. 6d. book. Next, it is advisable to learn something about the occurrence and appearance of the valuable minerals and the formations in which they are found. For all practical purposes I can recommend Cox and Ratte's "Mines and Minerals," one of the Technical Education series of New South Wales, which deals largely with the subject from an Australian standpoint, and is therefore particularly valuable to the Australian miner, but which will be found applicable to most other gold-bearing countries. I must not, however, omit to mention an admirably compiled *multum in parvo* volume prepared by Mr. G. Goyder, jun., Government Assayer and Assay Instructor at the School of Mines, Adelaide. It is called the "Prospectors' Pocketbook," costs only one shilling, is well bound, and of handy size to carry. In brief, plain language it describes how a man, having learned a little of assaying, may cheaply provide himself with a portable assay plant, and fluxes, and also gives considerable general information on the subject of minerals, their occurrence and treatment.[*]

[*] Another excellent and really practical book is Prof. Cole's "Practical Aids in Geology" (second edition), 10s. 6d.

It may here be stated that some twelve years ago I did a large amount of practical silver assaying on the Barrier (Broken Hill), which was not then so accessible a place as it is now, and got closely correct results from a number of different mines, with an extemporised plant almost amusing in its simplicity. All I took from Adelaide were a small set of scales capable of determining the weight of a button down to 20 ozs. to the ton, a piece of cheese cloth to make a screen or sieve, a tin ring 1 1/2 in. diameter, by 1/2 in. high, a small brass door knob to use as a cupel mould, and some powdered borax, carbonate of soda, and argol for fluxes; while for reducing lead I had recourse to the lining of a tea-chest, which lead contains no silver—John Chinaman takes good care of that. My mortar was a jam tin, without top or bottom, placed on an anvil; the pestle a short steel drill. The blacksmith at Mundi Mundi Station made me a small wrought iron crucible, also a pair of bent tongs from a piece of fencing-wire. The manager gave me a small common red flower pot for a muffle, and with the smith's forge (the fire built round with a few blocks of talcose schist) for a furnace, my plant was complete. I burned and crushed bones to make my bone-dust for cupelling, and thus provided made nearly forty assays, some of which were afterwards checked in Adelaide, in each instance coming as close as check assays generally do. Nowadays one can purchase cheaply a very effective portable plant, or after a few lessons a man may by practice make himself so proficient with the blowpipe as to obtain assay results sufficiently accurate for most practical purposes.

Coming then to the actual work of prospecting. What the prospector requires to know is, first, the usual locality of occurrence of the more valuable minerals; secondly, their appearance; thirdly, a simple mode of testing. With respect to occurrence, the older sandy and clay slates, chlorite slates, micaceous, and hornblendic schists, particularly at or near their junction with the intrusive granite and diorite, generally form the most likely geological country for the finding of mineral lodes, particularly gold, silver and tin. But those who have been engaged in practical mining for long, finding by experience that no two mineral fields are exactly alike in all their characteristics, have come to the conclusion that it is unwise to form theories as to why metals should or should not be found in certain enclosing rocks or matrices. Some of the best reef gold got in Victoria has been obtained in dead white, milky-looking quartz almost destitute of base metal. In South Australia reef gold is almost invariably associated with iron, either as oxide, as "gossan;" or ferruginous calcite, "limonite;" or granular silica, conglomerated by iron, the "ironstone" which forms the capping or outcrop of many of our reefs, and which is often rich in gold.

But to show that it is unsafe to decide off-hand in what class of matrix metals will or will not be found, I may say that in my own experience I have seen payable gold in the following materials:—

Quartz, dense and milky, also in quartz of nearly every colour and appearance, saccharoidal, crystalline, nay, even in clear glass-like six-sided prismatic crystals, and associated with silver, copper, lead, arsenic, iron as sulphide, oxide, carbonate, and tungstate, antimony, bismuth, nickel, zinc, lead, and other metals in one form or another; in slate, quartzite, mica schist, granite, diorite, porphyry, felsite, calcite, dolomite, common carbonate of iron, siliceous sinter from a hot spring, as at Mount Morgan; as alluvial gold in drifts formed of almost all these materials; and once, perhaps the most curious matrix of all, a small piece of apparently alluvial gold, naturally imbedded in a shaly piece of coal. This specimen, I think, is in the Sydney Museum. One thing, however, the prospector may make sure of: he will always find gold more or less intimately associated with silica (Quartz) in one or other of its many forms, just as he will always find cassiterite (oxide of tin) in the neighbourhood of granite containing muscovite (white mica), which so many people will persist in terming talc. It is stated to be a fact that tin has never been found more than about two miles from such granite.

From what has been said of its widely divergent occurrences it will be admitted that the Cornish miners' saying with regard to metals generally applies with great force to gold: "Where it is, there it is": and "Cousin Jack" adds, with pathetic emphasis, "and where it is generally, there I ain't."

I have already spoken of the geological "country rock" in which red gold is most likely to be discovered—i.e., the junction of the slates and schists with the igneous or metamorphic (altered) rocks, or in this vicinity. Old river beds formed of gravelly drifts in the same neighbourhood may probably contain alluvial gold, or shallow deposits of "wash" on hillsides and in valleys will often carry good surface gold. This is sometimes due to the denudation, or wearing away, of the hills containing quartz-veins—that is, where the alluvial gold really was derived from such veins, which, popular opinion to the contrary, is not always the case.

Much disappointment and loss of time and money may sometimes be prevented if prospectors will realise that *all* alluvial gold does not come from the quartz veins or reefs; and that following up an alluvial lead, no matter how rich, will not inevitably develop a payable gold lode. Sometimes gold, evidently of reef origin, is found in the alluvial; but in that case it is generally fine as regards the size of the particles, more or less sharp-edged, or crystalline in form if recently shed; while such gold is often of poorer quality than the true alluvial which occurs in mammillary (breast-like) nuggets, and is of a higher degree of purity as gold.

The ordinary non-scientific digger will do well to give credence to this view of the case, and will often thereby save himself much useless trouble. Sometimes also the alluvial gold, coarser in size than true reef-born alluvial, is derived almost *in situ* from small quartz "leaders," or veins, which the grinding down of the face of the slates has exposed; these leaders in time being also broken and worn, set free the gold they have contained, which does not, as a rule, travel far, but sometimes becomes water-worn by the rubbing over it of the disintegrated fragments of rock.

But the heavy, true alluvial gold, in great pure masses, mammillary, or botryoidal (like a bunch of grapes) in shape, have assuredly been formed by accretion on some metallic base, from gold salts in solution, probably chloride, but possibly sulphide.

Nuggets, properly so-called, are never found in quartz lodes; but, as will be shown later, a true nugget having all the characteristics of so-called water-worn alluvial may be artificially formed on a small piece of galena, or pyrites, by simply suspending the base metal by a thread in a vessel containing a weak solution of chloride of gold in which a few hard-wood chips are thrown.

Prospecting for alluvial gold at shallow depths is a comparatively easy process, requiring no great amount of technical knowledge. Usually the first gold is got at or near the surface and then traced to deep leads, if such exist.

At Mount Brown Goldfield, N.S.W., in 1881, I saw claimholders turning out to work equipped only with a small broom made of twigs and a tin dish. With the broom they carefully swept out the crevices of the decomposed slate as it was exposed on the surface, and putting the resulting dust and fragments into the tin dish proceeded to dry blow it.

The *modus operandi* is as follows: The operator takes the dish about half full of dirt, and standing with his back or side to the wind, if there be any, begins throwing the stuff up and catching it, or sometimes slowly pouring it from one dish to another, the wind in either case carrying away the finer particles. He then proceeds to reduce the quantity by carefully extracting the larger fragments of rock, till eventually he has only a handful or so of moderately fine "dirt" which contains any gold there may be. If in good sized nuggets it is picked out, if in smaller pieces or fine grains the digger slowly blows the sand and dust aside with his breath, leaving the gold exposed. This process is both tedious and unhealthy, and of course can only be carried out with very dry surface dirt. The stuff in which the gold occurred at Mount Brown was composed of broken slate with a few angular fragments of quartz. Yet, strange to say, the gold was invariably waterworn in appearance.

Dry blowing is now much in vogue on the West Australian fields owing to the scarcity of water; but the great objection is first, the large amount of dust the unfortunate dry blower has to carry about his person, and secondly, that the peck of dirt which is supposed to last most men a life time has to be made a continuous meal of every day.

For wet alluvial prospecting the appliances, besides pick and shovel, are puddling tub, tin dish, and cradle; the latter, a man handy with tools can easily make for himself.

In sinking, the digger should be careful to avoid making his shaft inconveniently small, and not to waste his energy by sinking a large "new chum" hole, which usually starts by being about three times too large for the requirements at the surface, but narrows in like a funnel at 10 feet or less. A shaft, say 4 feet by 2 feet 6 inches and sunk plumb, the ends being half rounded, is large enough for all requirements to a considerable depth, though I have seen smart men, when they were in a hurry to reach the drift, get down in a shaft even less in size.

The novice who is trying to follow or to find a deep lead must fully understand that the present bed of the surface river may not, in fact seldom does, indicate the ancient watercourses long since buried either by volcanic or diluvial action, which contain the rich auriferous deposits for which he is seeking; and much judgment and considerable underground exploration are often required to decide on the true course of leads. Only by a careful consideration of all the geological surroundings can an approximate idea be obtained from surface inspection alone; and the whole probable conditions which led to the present contour of the country must be carefully taken into account.

How am I to know the true bottom when I see it? asks the inexperienced digger. Well, nothing but long experience and intelligent observation will prevent mistakes at times, particularly in deep ground; but as a general rule, though it may sound paradoxical, you may know the bottom by the top.

That is, we will assume you are sinking in, say, 10 to 12 feet ground in a gully on the bank of which the country rock is exposed, and is, say, for instance, a clay slate or sandy slate set at a certain angle; then, in all probability, unless there be a distinct fault or change in the country rock between the slate outcrop and your shaft, the bottom will be a similar slate, standing at the same angle; and this will very probably be overlaid by a deposit of pipeclay, formed by the decomposition of the slates.

From the crevices of these slates, sometimes penetrating to a considerable distance, you may get gold, but it is useless attempting to sink through them. If the outcropping strata be a soft calcareous (limy) sandstone or soft felspathic rock, and that be also the true bottom, great care should be exercised or one is apt to sink through the bottom, which may be very loose and decomposed. I have known mistakes made in this way when many feet have been sunk, and driven through what was actually bed rock, though so soft as to deceive even men of experience. The formation, however, must be the guide, and except in some specially difficult cases, a man can soon tell when he is really on bed rock or "bottom."

On an alluvial lead the object of every one is to "get on the gutter," that is, to reach the lowest part of the old underground watercourse, through which for centuries the gold may have been accretionising from the percolation of the mineral-impregnated water; or, when derived from reefs or broken down leaders, the flow of water has acted as a natural sluice wherein the gold is therefore most thickly collected. Sometimes the lead runs for miles and is of considerable width, at others it is irregular, and the gold-bearing "gutter" small and hard to find. In many instances, for reasons not readily apparent, the best gold is not found exactly at the lowest portion of these narrow gutters, but a little way up the sides. This fact should be taken into consideration in prospecting new ground, for many times a claim has been deserted after cleaning up the "bottom," and another man has got far better gold considerably higher up on the sides of the gutter. For shallow alluvial deposits, where a man quickly works out his 30 by 30 feet claim, it may be cheaper at times to "paddock" the whole ground—that is, take all away from surface to bottom, but if he is in wet ground and he has to drive, great care should be taken to properly secure the roof by means of timber. How this may best be done the local circumstances only can decide.

CHAPTER III

LODE OR REEF PROSPECTING

The preceding chapter dealt more especially with prospecting as carried on in alluvial fields. I shall now treat of preliminary mining on lodes or "reefs."

As has already been stated, the likeliest localities for the occurrence of metalliferous deposits are at or near the junction of the older sedimentary formations with the igneous or intrusive rocks, such as granites, diorites, etc. In searching for payable lodes, whether of gold, silver, copper, or even tin in some forms of occurrence, the indications are often very similar. The first prospecting is usually done on the hilltops or ridges, because, owing to denudation by ice or water which have bared the bedrock, the outcrops are there more exposed, and thence the lodes are followed down through the alluvial covered plains, partly by their "strike" or "trend," and sometimes by other indicating evidences, which the practical miner has learned to know.

For instance, a lesson in tracing the lode in a grass covered country was taught me many years ago by an old prospector who had struck good gold in the reef at a point some distance to the east of what had been considered the true course. I asked him why he had opened the ground in that particular place. Said he, "Some folks don't use their eyes. You stand here and look towards that claim on the rise where the reef was last struck. Now, don't you see there is almost a track betwixt here and there where the grass and herbage is more withered than on either side? Why? Well, because the hard quartz lode is close to the surface all the way, and there is no great depth of soil to hold the moisture and make the grass grow."

I have found this simple lesson in practical prospecting of use since. But the strike or course of a quartz reef is more often indicated by outcrops, either of the silica itself or ironstone "blows," as the miners call them, but the term is a misnomer, as it argues the easily disproved igneous theory of veins of ejection, meaning thereby that the quartz with its metalliferous contents was thrown out in a molten state from the interior of the earth. This has in no case occurred, and the theory is an impossible one. True lodes are veins of injection formed by the infiltration of silicated waters carrying the metals also in solution. This water filled the fissures caused either by the cooling of the earth's crust, or formed by sudden upheavals of the igneous rocks.

Sometimes in alluvial ground the trend of the reef will be revealed by a track of quartz fragments, more or less thickly distributed on the surface and through the superincumbent soil. Follow these along, and at some point, if the lode be continuous, a portion of its solid mass will generally be found to protrude and can then again be prospected.

There is no rule as to the trend or strike of lodes, except that a greater number are found taking a northerly and southerly course than one which is easterly and westerly. At all events, such is the case in Australia, but it cannot be said that either has the advantage in being more productive. Some of the richest mines in Australasia have been in lodes running easterly and westerly, while gold, tin, and copper, in great quantity and of high percentage to the ton, have been got in such mines as Mount Morgan, Mount Bischoff, and the Burra, where there are no lodes properly so-called at all.

Mount Morgan is the richest and most productive gold mine in Australasia and amongst the best in the world.

Its yield for 1895 was 128,699 oz. of gold, valued at 528,700 pounds. Dividends paid in 1895, 300,000 pounds.

This mine was opened in 1886. Up to May 31, 1897, the total yield was 1,631,981 ozs. of gold, sold at 6,712,187 pounds, from which 4,400,000 pounds have been paid in dividends. (See *Mining Journal*, for Oct. 9, 1897.)

Mount Morgan shareholders have, in other words, divided over 43 1/2 tons of standard gold.

The Burra Burra Mine, about 100 miles from Adelaide, in a direction a little to the east of north, was found in 1845 by a shepherd named Pickett. It is singularly situated on bald hills standing 130 feet above the surrounding country. The ores obtained from this copper mine had been chiefly red oxides, very rich blue and green carbonates, including malachite, and also native copper. The discovery of this mine, supporting, as it did at one time, a large population, marked a new era in the history of the colony. The capital invested in it was 12,320 pounds in 5 pound shares, and no subsequent call was ever made upon the shareholders. The total amount paid in dividends was 800,000 pounds. After being worked by the original owners for some years the mine was sold to a new company, but during the last few years it has not been worked, owing in some degree to the low price of copper and also to the fact that the deposit then being worked apparently became exhausted. For many years the average yield was from 10,000 to 13,000 tons of ore, averaging 22 to 23 per cent of copper. It is stated that, during the twenty-nine and a half years in which the mine was worked, the company expended 2,241,167 in general expenses. The output of ore during the same period amounted to 234,648 tons, equal to 51,622 tons of copper. This, at the average price of copper, amounted to a money value of 4,749,224 pounds. The mine stopped working in 1877.

Mount Bischoff, Tasmania, has produced, since the formation of the Company to December 1895, 47,263 tons of tin ore. It is still in full work and likely to be for years to come.

Each of these immense metalliferous deposits was found outcropping on the summit of a hill of comparatively low altitude. There are no true walls nor can the ore be traced away from the hill in lode form. These occurrences are generally held to be due to hydrothermal or geyser action.

Then again lodes are often very erratic in their course. Slides and faults throw them far from their true line, and sometimes the lode is represented by a number of lenticular (double-pointed in section) masses of quartz of greater or less length, either continuing point to point or overlapping, "splicing," as the miners call it. Such formations are very common in West Australia. All this has to be considered and taken into account when tracing the run of stone.

This tyro also must carefully remember that in rough country where the lode strikes across hills and valleys, the line of the cap or outcrop will apparently be very sinuous owing to the rises and depressions of the surface. Many people even now do not understand that true lodes or reefs are portions of rock or material differing from the surrounding and enclosing strata, and continuing down to unknown depths at varying angles. Therefore, if you have a north and south lode outcropping on a hill and crossing an east and west valley, the said lode, underlying east, when you have traced its outcrop to the lowest point in the valley, between the two hills, will be found to be a greater or less distance, according to the angle of its dip or underlie, to the east of the outcrop on the hill where it was first seen. If it be followed up the next hill it will come again to the west, the amount of apparent deviation being regulated by the height of the hills and depth of the valley.

A simple demonstration will make this plain. Take a piece of half-inch pine board, 2 ft. long and 9 in. wide, and imagine this to be a lode; now cut a half circle out of it from the upper edge with a fret saw and lean the board say at an angle of 45 degrees to the left, look along the top edge, which you are to consider as the outcrop on the high ground, the bottom of the cut being the outcrop in the valley, and it will be seen that the lowest portion of the cut is some inches to the right; so it is with the lode, and in rough country very nice judgment is required to trace the true course.

For indications, never pass an ironstone "blow" without examination. Remember the pregnant Cornish saying with regard to mining and the current aphorism, "The iron hat covers the golden head." "Cousin Jack," put it "Iron rides a good horse." The ironstone outcrop may cover a gold, silver, copper or tin lode.

If you are searching for gold, the presence of the royal metal should be apparent on trial with the pestle and mortar; if silver, either by sight in one of its various forms or by assay, blowpipe or otherwise; copper will reveal itself by its peculiar colour, green or blue carbonates, red oxides, or metallic copper. It is an easy metal to prospect for, and its percentage is not difficult to determine approximately. Tin is more difficult to identify, as it varies so greatly in appearance.

Having found your lode and ascertained its course, you want next to ascertain its value. As a rule (and one which it will be well to remember) if you cannot find payable metal, particularly in gold "reef" prospecting, at or near the surface, it is not worth while to sink, unless, of course, you design to strike a shoot of metal which some one has prospected before you. The idea is exploded that auriferous lodes necessarily improve in value with depth. The fact is that the metal in any lode is not, as a rule, equally continuous in any direction, but occurs in shoots dipping at various angles in the length of the lode, in bunches or sometimes in horizontal layers. Nothing but actual exploiting with pick, powder, and brains, particularly brains, will determine this point.

Where there are several parallel lodes and a rich shoot has been found in one and the length of the payable ore ascertained, the neighbouring lodes should be carefully prospected opposite to the rich spot, as often similar valuable deposits will thus be found. Having ascertained that you have, say, a gold reef payable at surface and for a reasonable distance along its course, you next want to ascertain its underlie or dip, and how far the payable gold goes down.

As a general rule in many parts of Australia—though by no means an inflexible rule—a reef running east of north and west of south will underlie east; if west of north and east of south it will go down to the westward and so round the points of the compass till you come to east and west; when if the strike of the lodes in the neighbourhood has come round from north-east to east and west the underlie will be to the south; if the contrary was the case, to the north. It is surprising how often this mode of occurrence will be found to obtain. But I cannot too strongly caution the prospector not to trust to theory but to prove his lode and his metal by following it down on the underlie. "Stick to your gold" is an excellent motto. As a general thing it is only when the lode has been proved by an underlie shaft to water level and explored by driving on its course for a reasonable distance that one need begin to think of vertical shafts and the scientific laying out of the mine.

A first prospecting shaft need not usually be more than 5 ft. by 3 ft. or even 5 ft. by 2 ft. 6 in., particularly in dry country. One may often see in hard country stupid fellows wasting time, labour, and explosives in sinking huge excavations as much as 10 ft. by 8 ft. in solid rock, sometimes following down 6 inches of quartz.

When your shaft is sunk a few feet, you should begin to log up the top for at least 3 ft. or 4 ft., so as to get a tip for your "mullock" and lode stuff. This is done by getting a number of logs, say 6 inches diameter, lay one 7 ft. log on each side of your shaft, cut two notches in it 6 ft. apart opposite the ends of the shaft, lay across it a 5 ft. log similarly notched, so making a frame like a large Oxford picture frame. Continue this by piling one set above another till the desired height is attained, and on the top construct a rough platform and erect your windlass. If you have an iron handle and axle I need not tell you how to set up a windlass, but where timber is scarce you may put together the winding appliance described in the chapter headed "Rules of Thumb."

If you have "struck it rich" you will have the pleasure of seeing your primitive windlass grow to a "whip," a "whim," and eventually to a big powerful engine, with its huge drum and Eiffel tower-like "poppet heads," or "derrick," with their great spindle pulley wheels revolving at dizzy speed high in air.

"How shall I know if I have payable gold so as to save time and trouble in sinking?" says the novice. Truly it is a most important part of the prospector's art, whether he be searching for alluvial or reef gold, stream or lode tin, copper, or other valuable metal.

I presume you know gold when you see it?

If you don't, and the doubtful particle is coarse enough, take a needle and stick the point into the questionable specimen. If gold the steel point will readily prick it; if pyrites or yellow mica the point will glance off or only scratch it.

The great importance of the first prospect from the reef is well shown by the breathless intensity with which the two bearded, bronzed pioneer prospectors in some trackless Australian wild bend over the pan in which the senior "mate" is slowly reducing the sample of powdered lode stuff. How eagerly they examine the last pinch of "black sand" in the corner of the dish. Prosperity and easy times, or poverty and more "hard graft" shall shortly be revealed in the last dexterous turn of the pan. Let us hope it is a "pay prospect."

The learner, if he be far afield and without appliances of any kind, can only guess his prospect. An old prospector will judge from six ounces of stuff within a few pennyweights what will be the yield of a ton. I have seen many a good prospect broken with the head of a pick and panned in a shovel, but for reef prospecting you should have a pestle and mortar. The handiest for travelling is a mortar made from a mercury bottle cut in half, and a not too heavy wrought iron pestle with a hardened face. To be particular you require a fine screen in order to get your stuff to regulated fineness. The best for the prospector, who is often on the move, is made from a piece of cheesecloth stretched over a small hoop.

If you would be more particular take a small spring balance or an improvised scale, such as is described in Mr. Goyder's excellent little book, p. 14, which will enable you to weigh down to one-thousandth of a grain. It is often desirable to burn your stone before crushing, as it is thus more easily triturated and will reveal all its gold; but remember, that if it originally contained much pyrites, unless a similar course is adopted when treated in the battery, some of the gold will be lost in the pyrites.

Having crushed your gangue to a fine powder you proceed to pan it off in a similar manner to that of washing out alluvial earth, except that in prospecting quartz one has to be much more particular, as the gold is usually finer. The pan is taken in both hands, and enough water to cover the prospect by a few inches is admitted. The whole is then swirled round, and the dirty water poured off from time to time till the residue is clean quartz sand and heavy metal. Then the pan is gently tipped, and a side to side motion is given to it, thus causing the heavier contents to settle down in the corner. Next the water is carefully lapped in over the side, the pan being now tilted at a greater angle until the lighter particles are all washed away. The pan is then once more righted, and very little water is passed over the pinch of heavy mineral a few times, when the gold will be revealed in a streak along the bottom. In this operation, as in all others, only practice will make perfect, and a few practical lessons are worth whole pages of written instruction.

To make an amalgamating assay that will prove the amount of gold which can be got from a ton of your lode, take a number of samples from different parts, both length and breadth. The drillings from the blasting bore-holes collected make the best test. When finely triturated weigh off one or two pounds, place in a black iron pan (it must not be tinned), with 4 ozs. of mercury, 4 ozs. salt, 4 ozs. soda, and about half a gallon of boiling water; then, with a stick, stir the pulp constantly, occasionally swirling the dish as in panning off, till you feel certain that every particle of the gangue has come in contact with the mercury; then carefully pan off into another dish so as to lose no mercury. Having got your amalgam clean squeeze it through a piece of chamois leather, though a good quality of new calico previously wetted will do as well. The resulting pill of hard amalgam can then be wrapped in a piece of brown paper, placed on an old shovel, and the mercury driven off over a hot fire; or a clay tobacco pipe, the mouth being stopped with clay, makes a good retort (see "Rules of Thumb," pipe and potato retorting). The residue will be retorted gold, which, on being weighed and the result multiplied by 2240 for a 1 lb. assay, or by 1120 for 2 lb., will give the amount of gold per ton which an ordinary battery might be expected to save. Thus 1 grain to the pound, 2240 lbs. to the ton, would show that the stuff contained 4 oz. 13 dwt. 8 gr. per ton.

If there should be much base metal in your sample such as say stibnite (sulphide of antimony), a most troublesome combination to the amalgamator—instead of the formula mentioned above add to your mercury about one dwt. of zinc shavings or clippings, and to your water sufficient sulphuric acid to bring it to about the strength of vinegar (weaker, if anything, not stronger), place your material preferably in an earthenware or enamelled basin if procurable, but iron will do, and intimately mix by stirring and shaking till all particles have had an opportunity to combine with the mercury. Retort as before described. This device is my own invention.

The only genuine test after all is the battery, and that, owing to various causes, is often by no means satisfactory. First, there is a strong, almost unconquerable temptation to select the stone, thus making the testing of a few tons give an unduly high average; but more often the trouble is the other way. The stuff is sent to be treated at some inefficient battery with worn-out boxes, shaky foundations, and uneven tables, sometimes with the plates not half amalgamated, or coated with impurities, the whole concern superintended by a man who knows as little about the treatment of auriferous quartz by the amalgamating or any other processes as a dingo does of the differential calculus. Result: 3 dwt. to the ton in the retort, 30 dwt. in the tailings, and a payable claim declared a "duffer."

When the lode is really rich, particularly if it be carrying coarse gold, and owing to rough country, or distance, a good battery is not available, excellent results in a small way may be obtained by the somewhat laborious, but simple, process of "dollying." A dolly is a one man power single stamp battery, or rather an extra sized pestle and mortar (see "Rules of Thumb").

Silver lodes and lodes which frequently carry more or less gold, are often found beneath the dark ironstone "blows," composed of conglomerates held together by ferric and manganic oxides; or, where the ore is galena, the surface indications will frequently be a whitish limey track sometimes extending for miles, and nodules or "slugs" of that ore will generally be found on the surface from place to place. Most silver ores are easily recognisable, and readily tested by means of the blowpipe or simple fire assay. Sometimes the silver on being tested is found to contain a considerable percentage of gold as in the great Comstock lode in Nevada. Ore from the big Broken Hill silver load, New South Wales, also contains an appreciable quantity of the more precious metal. A natural alloy of gold containing 20 per cent silver, termed electrum, is the lowest grade of the noble metal.

Tin, lode, and stream, or alluvial, occurs only as an oxide, termed cassiterite, and yet you can well appreciate the compliment one Cornish miner pays to another whose cleverness he wishes to commend, when he says of him, "Aw, he do know tin," when you look at a representative collection of tin ores. In various shapes, from sharp-edged crystals to mammillary-shaped nuggets of wood-tin; from masses of 30 lbs. weight to a fine sand, like gunpowder, in colour black, brown, grey, yellow, red, ruby, white, and sometimes a mingling of several colours, it does require much judgment to know tin.

Stream tin is generally associated with alluvial gold. When such is the case there is no difficulty in saving the gold if you save the tin, for the yellow metal is of much greater specific gravity. As the natural tin is an oxide, and therefore not susceptible to amalgamation, the gold can be readily separated by means of mercury.

Lode tin sometimes occurs in similar quartz veins to those in which gold is got, and is occasionally associated with gold. Tin is also found, as at Eurieowie, in dykes, composed of quartz crystals and large scales of white mica, traversing the older slates. A similar occurrence takes place at Mount Shoobridge and at Bynoe Harbour, in the Northern Territory of South Australia; indeed, one could not readily separate the stone from these three places if it were mixed. As before stated tin will never be found far from granite, and that granite must have white mica as one of its constituents. It is seldom found in the darker coloured rocks, or in limestone country, but it sometimes occurs in gneiss, mica schist, and chlorite schist. Numerous other minerals are at times mistaken for tin, the most common of which are tourmaline or schorl, garnet, wolfram (which is a tungstate of iron with manganese), rutile or titanic acid, blackjack or zinc blende, together with magnetic, titanic, and specular iron in fine grains.

This rough and ready mode of determining whether the ore is tin is by weight and by scratching or crushing, when, what is called the "streak" is obtained. The colour of the tin streak is whitey-grey, which, when once known, is not easily mistaken. The specific gravity is about 7.0. Wolfram, which is most like it, is a little heavier, from 7.0 to 7.5, but its streak is red, brown, or blackish-brown. Rutile is much lighter, 4.2, and the streak light-brown; tourmaline is only 3.2. Blackjack is 4.3, and its streak yellowish-white.

I have seen several pounds weight to the dish got in some of the New South Wales shallow sinking tin-fields, and, as a rule, payable gold was also present. Fourteen years ago I told Western Australian people, when on a visit to that colony, that the neighbourhood of the Darling range would produce rich tin. Lately this had been proved to be the case, and I look forward to a great development of the tin mining industry in the south-western portion of Westralia.

The tin "wash" in question may also contain gold, as the country rock of the neighbourhood is such as gold is usually found in.[*]

> [*] Since this book was in the printers' hands, the discovery of payable gold has been reported from this district. A detailed discussion of methods of prospecting will be found in chapter ii. Of Le Neve Foster's "Ore and Stone Mining," and Mr. S. Herbert Cox's "Handbook for Prospectors."

CHAPTER IV

THE GENESIOLOGY OF GOLD—AURIFEROUS LODES

Up to a comparatively recent time it was considered heretical for any one to advance the theory that gold had been deposited where found by any other agency than that of fire. As late as 1860 Mr. Henry Rosales convinced himself, and apparently the Victorian Government also, that quartz veins with their enclosed metal had been ejected from the interior of the earth in a molten state. His essay, which is very ingenious and cleverly written, obtained a prize which the Government had offered, but probably Mr. Rosales himself would not adduce the same arguments in support of the volcanic or igneous theory to-day. His phraseology is very technical; so much so that the ordinary inquirer will find it somewhat difficult to follow his reasoning or understand his arguments, which have apparently been founded only on the occurrence of gold in some of the earlier discovered quartz lodes, and the conclusions at which he arrived are not borne out by later experience. He says:— "While, however, there are not apparent signs of mechanical disturbances, during the long period that elapsed from the cooling of the earth's surface to the deposition of the Silurian and Cambrian systems, it is to be presumed that the internal igneous activity of the earth's crust was in full force, so that on the inner side of it, in obedience to the laws of specific gravity, chemical attraction, and centrifugal force, a great segregation of silica in a molten state took place. This molten silica continually accumulating, spreading, and pressing against the horizontal Cambro-Silurian beds during a long period at length forced its way through the superincumbent strata in all directions; and it is abundantly evident, under the conditions of this force and the resistance offered to its action, that the line it would and must choose would be along any continuous and slightly inclined diagonal, at times crossing the strata of the schists, though generally preferring to develop itself and egress between the cleavage planes and dividing seams of the different schistose beds."

He goes on to say, "Another argument to the same end (i.e., the igneous origin) may be shown from the fact that the auriferous quartz lodes have exercised a manifest metamorphic action on the adjacent walls or casing; they have done so partly in a mineralogical sense, but generally there has been a metamorphic alteration of the rock." Mr. Rosales then tells his readers, what we all know must be the case, that the gold would be volatilised by the heat, as would be also the other metals, which he says, were in the form of arseniurets and sulphurets; but he fails to explain how the sublimated metals afterwards reassumed their metallic form. Seeing that, in most cases, they would be hermetically enclosed in molten and quickly solidifying silica they could not be acted on to any great extent by aqueous agency. Neither does Mr. Rosales's theory account at all for auriferous lodes; which below water level are composed of a solid mass of sulphide of iron with traces of other sulphides, gold, calcspar, and a comparatively small percentage of silica. Nor will it satisfactorily explain the auriferous antimonial silica veins of the New England district, New South Wales, in which quantities of angular and unaltered fragments of slate from the enclosing rocks are found imbedded in the quartz.

With respect to the metamorphism of the enclosing rocks to a greater degree of hardness, which Mr. Rosales considered was due to heat, it should be remembered that these rocks in their original state were much softer and more readily fusible than the quartz, consequently all would have been molten and mingled together instead of showing as a rule clearly defined walls. It is much more rational to suppose that the increased hardness imparted to the slates and schists at or near their contact with the lode is due to an infiltration of silica from the silicated solution which at one time filled the fissure. Few scientists can now be found to advance the purely igneous theory of lode formation, though it must be admitted that volcanic action has probably had much influence not only in the formation of mineral veins, but also on the occurrence of the minerals therein. But the action was hydrothermal, just such as was seen in course of operation in New Zealand a few years ago when, in the Rotomahana district, one could actually see the growing of the marvellous White and Pink Terraces formed by the release of silica from the boiling water exuding from the hot springs, which water, so soon as the heat and pressure were removed, began to deposit its silica very rapidly; while at the Thames Gold-field, in the same country hot, silicated water continuously boiled out of the walls of some of the lodes after the quartz had been removed and re-deposited a siliceous sinter thereon.

On this subject I note the recently published opinions of Professor Lobley, a gentleman whose scientific reputation entitles his utterances to respect, but who, when he contends that gold is not found in the products of volcanic action is, I venture to think, arguing from insufficient premises. Certainly his theories do not hold good either in Australasia or America where gold is often, nay, more usually, found at, or near, either present or past regions of volcanic action.

It is always gratifying to have one's theories confirmed by men whose opinions carry weight in the scientific world. About seventeen years ago I first published certain theories on gold deposition, which, even then, were held by many practical men, and some scientists, to be open to question. Of late years, however, the theory of gold occurrence by deposition from mineral salts has been accepted by all but the "mining experts" who infest and afflict the gold mining camps of the world. These opine that gold ought to occur in "pockets" only (meaning thereby their own).

Recently Professor Joseph Le Conte, at a meeting of the American Institute of Mining Engineers, criticised a notable essay on the "Genesis of Ore Deposits," by Bergrath F. Posepny. The Professor's general conclusions are:

1. "Ore deposits, using the term in its widest sense, may take place from any kinds of waters, but especially from alkaline solutions, for these are the natural solvents of metallic sulphides, and metallic sulphides are usually the original form of such deposits."

2. "They may take place from waters at any temperature and any pressure, but mainly from those at high temperature and under heavy pressure, because, on account of their great solvent power, such waters are heavily freighted with metals."

3. "The depositing waters may be moving in any direction, up-coming, horizontally moving, or even sometimes down-going, but mainly up-coming; because by losing heat and pressure at every step such waters are sure to deposit abundantly."

4. "Deposits may take place in any kind of waterways—in open fissures, in incipient fissures, joints, cracks, and even in porous sandstone, but especially in great open fissures, because these are the main highways of ascending waters from the greatest depths."

5. "Deposits may be found in many regions and in many kinds of rocks, but mainly in mountain regions, and in metamorphic and igneous rocks, because the thermosphere is nearer the surface, and ready access thereto through great fissures is found mostly in these regions and in these rocks."

These views are in accordance with nearly all modern research into this interesting and fruitful subject.

Among the theories which they discredit is that ore bodies may usually be assumed to become richer in depth. As applied to gold lodes the teaching of experience does not bear out this view.

If it be taken into account that the time in which most of our auriferous siliceous lodes were formed was probably that indicated in Genesis as before the first day or period when "the earth was without form and void, and darkness was upon the face of the deep," it will be realised that the action we behold now taking place in a small way in volcanic regions, was probably then almost universal. The crust of the earth had cooled sufficiently to permit water to lie on its surface, probably in hot shallow seas, like the late Lake Rotomahana. Plutonic action would be very general, and volcanic mud, ash, and sand would be ejected and spread far and wide, which, sinking to the bottom of the water, may possibly be the origin of what we now designate the azoic or metamorphic slates and schists, as also the early Cambrian and Silurian strata. These, from the superincumbent weight and internal heat, became compacted, and, in some cases, crystallised, while at the same time, from the ingress of the surface waters to the heated regions below, probably millions of geysers were spouting their mineral impregnated waters in all directions; and in places where the crust was thin, explosions of super-heated steam caused huge upheavals, rifts, and chasms, into which these waters returned, to be again ejected, or to be the cause of further explosions. Later, as the cooling-process continued, there would be shrinkages of the earth's crust causing other fissures; intrusive granites further dislocated and upheaved the slates. About this age, probably, when really dry land began to appear, came the first formation of mineral lodes, and the waters, heavily charged with silicates, carbonates of lime, sulphides, etc., in solution, commenced to deposit their contents in solid form when the heat and pressure were removed.

I am aware that part of the theory here propounded as to the probable mode of formation of the immense sedimentary beds of the Archaic or Azoic period is not altogether orthodox—i.e., that the origin of these beds is largely due to the ejection of mud, sand, and ashes from subterraneous sources, which, settling in shallow seas, were afterwards altered to their present form. It is difficult, however, to believe that at this very early period of geologic history so vast a time had elapsed as would be required to account for these enormous depositions of sediment, if they were the result only of the degradation of previously elevated portions of the earth's surface by water agency. Glacial action at that time would be out of the question.

But what about the metals? Whence came the metallic gold of our reefs and drifts? What was it originally—a metal or a metallic salt, and if the latter, what was its nature?—chloride, sulphide, or silicate, one, or all three? I incline to the latter hypothesis. All three are known, and the chemical conditions of the period were favorable for their natural production. Assuming that they did exist, the task of accounting for the mode of occurrence of our auriferous quartz lodes is comparatively simple. Chloride of gold is at present day contained in sea water and in some mineral waters, and would have been likely to be more abundant during the Azoic and early Paleozoic period.

Sulphide of gold would have been produced by the action of sulphuretted hydrogen; hence probably our auriferous pyrites lodes, while silicate of gold might have resulted from a combination of gold chlorides with silicic acid, and thus the frequent presence of gold in quartz is accounted for.

A highly interesting and instructive experiment, showing how gold might be, and probably was, deposited in quartz veins, was carried out by Professor Bischof some years ago. He, having prepared a solution of chloride of gold, added thereto a solution of silicate of potash, whereupon, as he states, the yellow colour of the chloride disappeared, and in half an hour the fluid turned blue, and a gelatinous dark-blue precipitate appeared and adhered to the sides of the vessel. In a few days moss-like forms were seen on the surface of the precipitate, presumably approximating to what we know as dendroidal gold—that is, having the appearance of moss, fern, or twigs. After allowing the precipitate to remain undisturbed under water for a month or two a decomposition took place, and in the silicate of gold specks of metallic gold appeared. From this, the Professor argues, and with good show of reason, that as we know now that the origin of our quartz lodes was the silicates contained in certain rocks, it is probable that a natural silicate of gold may be combined with these silicates. If this can be demonstrated, the reason for the almost universal occurrence of gold in quartz is made clear.

About 1870, Mr. Skey, analyst to the New Zealand Geological Survey Department, made a number of experiments of importance in respect to the occurrence of gold. These experiments were summarised by Sir James Hector in an address to the Wellington Philosophical Society in 1872. Mr. Skey's experiments disproved the view generally held that gold is unaffected by sulphur or sulphuretted hydrogen gas, and showed that these elements combined with avidity, and that the gold thus treated resisted amalgamation with mercury. Mr. Skey proved the act of absorption of sulphur by gold to be a chemical act, and that electricity was generated in sufficient quantity and intensity during the process to decompose metallic solutions. Sulphur in certain forms had long been known to exercise a prejudicial effect upon the amalgamation of gold, but this had always been attributed to the combination of the sulphur with the quicksilver used. Now, however, it is certain that the sulphurising of the gold must be taken into account. We must remember that the particles of gold in the stone may be enveloped with a film of auriferous sulphide, by which they are protected from the solvent actions of the mercury. The sulphurisation of the gold gives no ocular manifestation by change of colour or perceptible increase of weight, as in the case of the formation of sulphides of silver, lead and other metals, on account of the extremely superficial action of the sulphur, and hence probably the existence of the gold sulphide escaped detection by chemists.

Closely allied to this subject is the investigation of the mode in which certain metals are reduced from their solutions by metallic sulphides, or, in common language, the influence which the presence of such substances as mundic and galena may exercise in effecting the deposit of pure metals, such as gold, in mineral lodes. The close relation which the richness of gold veins bears to the prevalence of pyrites has been long familiar both to scientific observers and to practical miners. The gold is an after deposit to the pyrites, and, as Mr. Skey was the first to explain, due to its direct reducing influences. By a series of experiments Mr. Skey proved that the reduction of the metal was due to the direct action of the sulphide, and showed that each grain of iron pyrites, when thoroughly oxidised, will reduce 12 1/4 grains of gold from its solution as chloride. He also included salts of platina and silver in this general law, and demonstrated that solutions of any of these metals traversing a vein rock containing certain sulphides would be decomposed, and the pure metal deposited. We are thus enabled to comprehend the constant association of gold, or native alloys of gold and silver, in veins which traverse rocks containing an abundance of pyrites, whether they have been formed as the result of either sub-aqueous volcanic outburst or by the metamorphism of the deeper-seated strata which compose the superficial crust of the earth.

Mr. Skey also showed by very carefully conducted experiments that the metallic sulphides are not only better conductors of electricity than has hitherto been supposed, but that when paired they were capable of exhibiting strong electro-motive power. Thus, if galena and zinc blende in acid solutions be connected in the usual manner by a voltaic pair, sulphuretted hydrogen is evolved from the surface of the former, and a current generated which is sufficient to reduce gold, silver or copper from their solutions in coherent electro-plate films. The attributing of this property of generating voltaic currents, hitherto supposed to be almost peculiar to metals, to such sulphides as are commonly found in metalliferous veins, further led Mr. Skey to speculate how far the currents discovered to exist in such veins by Mr. E. F. Fox might be produced by the gradual oxidation of mixed sulphides, and that veins containing bands of different metallic sulphides, bounded by continuing walls, and saturated with mineral waters, may constitute under some circumstances a large voltaic battery competent to produce electro-deposition of metals, and that the order of the deposit of these mineral lodes will be found to bear a definite relation to the order in which the sulphides rank in the table of their electro-motive power. These researches may lead to some clearer comprehension of the law which regulates the distribution of auriferous veins, and may explain why in some cases the metal should be nearly pure, while in others it is so largely alloyed with silver.

The following extract was lately clipped from a mining paper. If true, the experiment is interesting:—

"An American scientist has just concluded a very interesting and suggestive experiment. He took a crushed sample of rich ore from Cripple Creek, which carried 1100 ozs. of gold per ton, and digested it in a very weak solution of sodium chloride and sulphate of iron, making the solution correspond as near as practicable to the waters found in Nature. The ore was kept in a place having a temperature little less than boiling water for six weeks, when all the gold, except one ounce per ton, was found to have gone into solution. A few small crystals of pyrite were then placed in the bottle of solution, and the gold began immediately to precipitate on them. It was noticeable, however, that the pyrite crystals which were free from zinc, galena, or other extraneous matter received no gold precipitate. Those which had such foreign associations were beautifully covered with fine gold crystals."

Experimenting in a somewhat similar direction abut twelve months since, I found that the West Australian mine water, with the addition of an acid, was a solvent of gold. The idea of boiling it did not occur to me, as the action was rapid in cold water.

Assuming, then, that gold originally existed as a mineral salt, when and how did it take metallic form? Doubtless, just in the same manner as we now (by means of well-known reagents which are common in nature) precipitate it in the laboratory. With regard to that found in quartz lodes finely disseminated through the gangue, the change was brought about by the same agency which caused the silicic acid to solidify and take the form in which we now see it in the quartz veins. Silica is soluble in solutions of alkaline carbonates, as shown in New Zealand geysers; the solvent action being increased by heat and pressure, so also would be the silicate or sulphide of gold. When, however, the waters with their contents were released from internal pressure and began to lose their heat the gold would be precipitated together with the salts of some other metals, and would, where the waters could percolate, begin to accretionise, thus forming the heavy or specimen gold of some reefs. On this class of deposition I shall have more to say when treating of the origin of alluvial gold in the form of nuggets.

Mr. G. F. Becker, of the United States Geological Survey, writing of the geology of the Comstock lode, says:—"Baron Von Richthofen was of opinion that fluorine and chlorine had played a large part of the ore deposition in the Comstock, and this the writer is not disposed to deny; but, on the other hand, it is plain that most of the phenomena are sufficiently accounted for on the supposition that the agents have been merely solutions of carbonic and hydrosulphuric acids. These reagents will attack the bisilicates and felspars. The result would be carbonates and sulphides of metals, earth, alkalies, and free quartz, but quartz and sulphides of the metals are soluble in solutions of carbonates and sulphides of the earths and alkalies, and the essential constituents of the ore might, therefore, readily be conveyed to openings in the vein where they would have been deposited on relief of pressure and diminution of temperature. An advance boring on the 3000 ft. level of the Yellow Jacket struck a powerful stream of water at 3065 ft. (in the west country), which was heavily charged with hydrogen sulphide, and had a temperature of 170 degrees F., and there is equal evidence of the presence of carbonic acid in the water of the lower levels. A spring on the 2700 ft. level of the Yellow Jacket which showed a temperature of about 150 degrees F., was found to be depositing a sinter largely composed of carbonates."

It may be worth while here to speak of the probable reason why gold, and indeed almost all the metals generally occur in shutes in the lodes; and why, as is often the case, these shutes are found to be more or less in a line with each other in parallel lodes, and why also the junction of two lodes is frequently specially productive. The theory with respect to these phenomena which appears most feasible is, that at these points certain chemical action has taken place, by which the deposition of the metals has been specially induced. Generally a careful examination of the enclosing rocks where the shute is found will reveal some points of difference from the enclosing rocks at other parts of the course of the lode, and when ore shutes are found parallel in reefs running on the same course, bands or belts of similar country rock will be found at the productive points. From this we may fairly reason that at these points the slow stream filling the lode cavity met with a reagent percolating from this particular band of rock, which caused the deposition of its metals; and, indeed, I am strongly disposed to believe that the deposition of metals, particularly in some loose lodes, may even now be proceeding. But as in Nature's laboratory the processes, if certain, are slow, this theory may be difficult to prove.

Why the junction of lodes is often found to be more richly metalliferous than neighbouring parts is probably because there the depositing reagents met. This theory is well put by Mr. S. Herbert Cox, late of Sydney, in his useful book, "Mines and Minerals." He says:—
"It is a well-known fact in all mining districts that the junctions of lodes are generally the richest points, always supposing that this junction takes place in 'kindly country,' and the explanation of this we think is simple on the aqueous theory of filling of lodes. The water which is traversing two different channels of necessity passes through different belts of country, and will thus have different minerals in solution. As a case in point, let us suppose that the water in one lode contained in solution carbonates of lime, and the alkalies and silica derived form a decomposition of felspars; and that the other, charged with hydro-sulphuric acid, brought with it sulphide of gold dissolved in sulphide of lime. The result of these two waters meeting would be that carbonate of lime would be formed, hydrosulphuric acid would be set free, and sulphide of gold would be deposited, as well as silica, which was formerly held in solution by the carbonic acid."

Most practical men who have given the subject attention will, I think, be disposed to coincide with this view, though there are some who hold that the occurrence of these parallel ore shutes and rich deposits at the junctions of lodes is due to extraneous electrical agency. Of this, however I have failed to find any satisfactory evidence.

There is, however, proof that lodes are actually re-forming and the action observed is very interesting as showing how the stratification in some lodes has come about. Instances are not wanting of the growth of silica on the sides of the drives in mines. This was so in some of the mines on the Thames, New Zealand, previously mentioned, where in some cases the deposition was so rapid as to be noticeable from day to day, whilst the big pump was actually choked by siliceous deposits. In old auriferous workings which have been under water for years, in many parts of the world, formations of iron and silica have been found on the walls and roof, while in mining tunnels which have been long unused stalactites composed of silica and calcite have formed. Then, again, experiments made by the late Professor Cosmo Newbery, in Victoria, showed that a distinctly appreciable amount of gold, iron, and silica (the latter in granular form) could be extracted from solid mine timber; which had been submerged for a considerable time.

This reaction then must be in progress at the present time, and doubtless under certain conditions pyrites would eventually take the place of the timber, as is the case with some of the long buried driftwood found in Victorian deep leads. Again, we know that the water from some copper mines is so charged with copper sulphate that if scrap iron be thrown into it, the iron will be taken up by the sulphuric acid, and metallic copper deposited in its place. All this tends to prove that the deposition of metals from their salts, though probably not now as rapid as formerly, is still ceaselessly going on in some place or another where the necessary conditions are favourable.

With regard to auriferous pyritic lodes, it does not appear even now to be clear, as some scientists assert, that their gold is never found in chemical combination with the sulphides of the base metals. On the contrary, I think much of the evidence points in the other direction.

I have long been of opinion that it is really so held in many of the ferro-sulphides and arsenio-ferro sulphides. On this subject Mr. T. Atherton contributed a short article in 1891 to the *Australian Mining*

Standard which is worthy of notice. He says, referring to an occurrence of a Natural Sulphide of Gold: "The existence of gold, in the form of a natural sulphide in conjunction with pyrites, has often been advanced theoretically, as a possible occurrence; but up to the present time has, I believe, never been established as an actual fact. During my investigation on the ore of the Deep Creek mines, Nambucca, New South Wales, I have found in them what I believe to be gold existing as a natural sulphide. The lode is a large irregular one of pure arsenical pyrites carrying, in addition to gold and silver, nickel and cobalt. It exists in a felsite dyke immediately on the coast. Surrounding it on all sides are micaceous schists, and in the neighbourhood about half a mile distant is a large granite hill about 800 feet high. In the lode and its walls are large quantities of pyrophyllite, and in some parts of the mine there are deposits of pure white translucent mica, but in the ore itself it is a yellow or pale olive green, and is never absent from the pyrites.

"From the first I was much struck with the exceedingly fine state of division in which the gold existed in the ore. After roasting and very carefully grinding down in an agate mortar, I have never been able to get any pieces of gold exceeding one-thousandth of an inch in diameter, and the greater quantity is very much finer than this. Careful dissolving of the pyrites and gangue so as to leave the gold intact failed to find it in any larger diameter. As this was a very unusual experience in investigations on many other kinds of pyrites, I was led further into the matter.

"Ultimately, after a number of experiments, there was nothing left but to test for gold existing as a natural sulphide. Taking 200 gr. of ore from a sample assaying 17 oz. fine gold per ton, grinding it finely and heating for some hours with yellow sodium sulphide—on decomposing the filtrate and treating for gold I got a result at the rate of 12 oz. per ton. This was repeated several times with the same result.

"This sample came from the lode at the 140 ft. level, whilst samples from the higher levels where the ore is more oxidised, although carrying the gold in exactly the same degree of fineness, do not give as high a percentage of auric sulphide.

"It would appear that all the gold in the pyrites (and I have never found any gold existing apart from the pyrites) has originally taken its place there as a sulphide."

Professor Newbery, who made many valuable suggestions on the subject, says, speaking of gold in pyritous lodes:

"As it (the gold salt) may have been in the same solution that deposited the pyrites, which probably contained its iron in the form of proto-carbonate with sulphates, it was not easy at first to imagine any ordinary salt of gold; but this I find can be accomplished with very dilute solutions in the presence of an alkaline carbonate and a large excess of carbonic acid, both of which are common constituents of mineral waters, especially in Victoria. This is true of chloride of gold, and if the sulphide is required in solution, it is only necessary to charge the solution with an excess of sulphuretted hydrogen. In this matter both sulphides may be retained in the same solution, depositing gradually with the escape of the carbonic acid."

Pyritic lodes usually contain a considerable proportion of calcareous matter, mostly carbonates, and consequently it appears not improbable that the gold may remain in some instances as a sulphide, particularly in samples of pyrites, in which it cannot be detected even by the microscope until by calcination the iron sulphide is changed to an oxide, wherein the gold may be seen in minute metallic specks. The whole subject is full of interest, and careful scientific investigation may lead to astonishing results.

CHAPTER V

THE GENESIOLOGY OF GOLD – AURIFEROUS DRIFTS

Having considered the origin of auriferous lodes, and the mode by which in all probability the gold was conveyed to them and deposited as a metal, it is necessary also to inquire into the derivation of the gold of our auriferous drifts, and the reasons for its occurrence therein.

When quite a lad on the Victorian alluvial fields, I frequently heard old diggers assert that gold grew in the drifts where found. At the time we understood this to mean that it grew like potatoes; and, although not prepared with a scientific argument to prove that such was not so, the idea was generally laughed at. I have lived to learn that these old hard-heads were nearer the truth than possibly they clearly realised, and that gold does actually grow or agglomerate; and, indeed, is probably even now thus growing, though it is likely that the chemical and electric action in the mineral waters flowing through the drifts is not in this age nearly so active as formerly.

Most boys have tried the experiment of dipping a clean-bladed knife into sulphate of copper, and so depositing on the steel a film of copper, which adheres closely until worn away. This is a simple demonstration of a hydro-metallurgical process, though probably young hopeful is not aware of the fact; and it is really by an enlargement of this process that our beautiful and artistic gold-and silver-plated ware is produced.

In the great laboratory of Nature similar chemical depositions have taken place in the past, and may still be in progress; indeed, there is sound scientific reason to suppose that in certain localities this is even now the case, and that in this way much of our so-called alluvial gold has been formed, that is, by the deposition on metallic bases of the gold held in solution.

We will, however, take, to begin with, the generally accepted theory as to the occurrence of alluvial gold. First, let it be said, that certain alluvial gold is unquestionably derived from the denudation of quartz lodes. Such is the gold dust found in many Asiatic and

African rivers, in the great placer mines of California, as also the gold dust gained from the beach sand on the west coast of New Zealand, or in the enormous alluvial drifts of the Shoalhaven Valley, New South Wales. Of the first, many fabulous tales are told to account for its being found in particular spots each summer after the winter floods, and miraculous agency was asserted, while the early beachcombers of the Hokitika district found an equally ridiculous derivation for their gold, which was always more plentiful after heavy weather. They imagined that the breakers were disintegrating some abnormally rich auriferous reefs out at sea, and that the resultant gold was washed up on the beach.

The facts are simply, with regard to the rivers, that the winter floods break down the drifts in the banks and agitate the auriferous detritus, thus acting as natural sluices, and cause the metal to accumulate in favourable spots; whilst on the New Zealand coast the heavy seas breaking on the shingly beach, carry off the lighter particles, leaving behind the gold, which is so much heavier. These beaches are composed, as also are the "terraces" behind, of enormous glacial and fluvial deposits, all containing more or less gold, and extend inland to the foot of the mountains.

It is almost certain that the usually fine gold got by hydraulicing in Californian canyons, in the gullies of the New Zealand Alps, and the great New South Wales drifts, is largely the result of the attrition of the boulders and gravel of moraines, which has thus freed, to a certain extent, the auriferous particles. But when we find large nuggety masses of high carat gold in the beds of dead rivers, another origin has to be sought.

As previously stated, there is fair reason to assume that at least three salts of gold have existed, and, possibly, may still be found in Nature—silicate, sulphide, and chloride. All of these are soluble and in the presence of certain reagents, also existing naturally, can be deposited in metallic form. Therefore, if, as is contended, reef gold was formed with the reefs from solutions in mineral waters, by inferential reasoning it can be shown that much of our alluvial gold was similarly derived.

The commonly accepted theory, however, is that the alluvial matter of our drifts has been ground out of the solid siliceous lodes by glacial and fluvial action, and that the auriferous leads have been formed by the natural sluicing operations of former streams. To this, however, there are several insuperable objections.

First, how comes it that alluvial gold is usually superior in purity to the "reef" gold immediately adjacent? Second, why is it that masses of gold, such as the huge nuggets found in Victoria and New South Wales, have never been discovered in lodes? Third, how is it that these heavy masses which, from their specific gravity, should be found only at the very bottom of the drifts, if placed by water action, are sometimes found in all positions from the surface to the bottom of the "wash"? And, lastly, why is it that when an alluvial lead is traced up to, or down from, an auriferous reef, that the light, angular gold lies close to the roof, while the heavy masses are often placed much farther away? Any one who has worked a ground sluice knows how extremely difficult it is with a strong head of water to shift from its position an ounce of solid gold. What, then, would be the force required to remove the Welcome Nugget? Under certain circumstances, Niagara would not be equal to the task.

The generally smooth appearance of alleged alluvial gold is adduced as an argument in favour of its having been carried by water from its original place of deposit, and thus in transit become waterworn; while some go so far as to say that it was shot out of the reefs in a molten state. The latter idea has been already disposed of, but if not, it may be dismissed with the statement that the heat which would melt silica in the masses met with in lodes would sublimate any gold contained, and dissipate it, not in nuggets but in fumes. With regard to the assumed waterworn appearance of alluvial gold, I have examined with the microscope the smooth surface of more than one apparently waterworn nugget, and found that it was not scratched and abraded, as would have been the case had it been really waterworn, but that it presented the same appearance, though infinitely finer in grain, as the surface of a piece of metal fresh from the electrical plating-bath.

Mr. Daintree, of the Victorian Geological Survey, many years ago discovered accidentally that gold chloride would deposit its metal on a metallic base in the presence of any organic substance. Mr.

Daintree found that a piece of undissolved gold in a bottle containing chloride of gold in solution had, owing to a portion of the cork having fallen into the liquid, grown or accretionised so much that it could not be extracted through the neck. This lead Mr. Charles Wilkinson, who has contributed much to our scientific knowledge of metallurgy, to experiment further in the same direction. He says: "Using the most convenient salt of gold, the terchloride, and employing wood as the decomposing agent, in order to imitate as closely as possible the organic matter supposed to decompose the solution circulating through the drifts, I first immersed a piece of cubic iron pyrites taken from the coal formation of Cape Otway, far distant from any of our gold rocks, and therefore less likely to contain gold than other pyrites. The specimen (No. 1) was kept in dilute solution for about three weeks, and is completely covered with a bright film of gold. I afterwards filed off the gold from one side of a cube crystal to show the pyrites itself and the thickness of the surrounding coating, which is thicker than ordinary notepaper. If the conditions had continued favourable for a very lengthened period, this specimen would doubtless have formed the nucleus of a large nugget. Iron, copper, and arsenical pyrites, antimony, galena, molybdenite, zinc blende, and wolfram were treated in the above manner with similar results. In the above experiments a small chip of wood was employed as the decomposing agent. In one instance I used a piece of leather. All through the wood and leather gold was disseminated in fine particles, and when cut through the characteristic metallic lustre was brightly reflected. The first six of these sulphides were also operated upon simply in the solution without organic matter; but they remained unaltered."

Wilkinson found that when the solution of gold chloride was as strong as, say, four grains to the ounce of water, that the pyrites or other base began to decompose, and the iron sulphide changed to yellow oxide, the "gossan" of our lodes, and that though the gold was deposited, this occurred in an irregular way, and it was coated with a dark brown powdery film something like the "black gold," often found in drifts containing much ferruginous matter. Such were the curious Victorian nuggets Spondulix and Lothair.

Professor Newbery also made a number of similar experiments, and arrived at like results. He states as follows: "I placed a cube of galena in a solution of chloride of gold, with free access of air, and put in organic matter; gold was deposited as usual, in a bright metallic film, apparently completely coating the cube. After a few months the film burst along the edges of the cube, and remained in that state with the cracks open without any further alteration in size or form being apparent. Upon removing it a few days ago and breaking it open, I found that a large portion of the galena had been decomposed, forming chloride and sulphate of lead and free sulphur, which were mixed together, encasing a small nucleus of undecomposed sulphate of lead. The formation of these salts had exerted sufficient force to burst open the gold coating, which upon the outside had the mammillary form noticed by Wilkinson, while the inside was rough and irregular with crystals forcing their way into the lead salts. Had this action continued undisturbed, the result would have been a nugget with a nucleus of lead salts, or if there had been a current to remove the results of decomposition, a nugget without a nucleus of foreign matter."

But Newbery also made another discovery which still further establishes the probability of the accretionary growth of gold in drifts. In the first experiments both investigators used organic substances as the reagent to cause the deposit of gold on its base, and in each case these substances whether woodchips, leather, or even dead flies, were found to be so absolutely impregnated with gold as to leave a golden skeleton when afterwards burned. Timber found in the Ballarat deep leads has been proved to be similarly impregnated.

Newbery found that gold could also be deposited on sulphurets without any other reagent. He says: "In our mineral sulphurets, however, we have agents which are not only capable of reducing gold and silver from solution, but besides are capable of locating them when so reduced in coherent and bulky masses. Thus the aggregation of the nuggety forms of gold from solution becomes a still more simple matter, only one reagent being necessary, so that there is a greater probability of such depositions obtaining than were a double process necessary. Knowing the action of sulphides, the manner or the mode of formation of a portion at least of these nuggets seems apparent. Conceive a stream or river fed by springs rising in a country intersected by auriferous reefs, and consequently in this case carrying gold in solution; the drift of such a country must be to a greater or lesser extent pyritous, so that the *debris* forming the beds of these streams or rivers will certainly contain nodules of such matters disseminated or even stopping them in actual contact with the flow of water. It follows, then, from what has been previously affirmed, that there will be a reduction of gold by these nodules, and that the metal thus reduced will be firmly attached to them, at first in minute spangles isolated from each other, but afterwards accumulating and connecting in a gradual manner at that point of the pyritous mass most subject to the current until a continuous film of some size appears. This being formed the pyrites and gold are to a certain extent polarised, the film or irregular but connected mass of gold forming the negative, and the pyrites the positive end of a voltaic pair; and so according as the polarisation is advanced to completion the further deposition of gold is changed in its manner from an indiscriminate to an orderly and selective deposition concentrated upon the negative or gold plate. The deposition of gold being thus controlled, its loss by dispersion or from the crumbling away of the sustaining pyrites is nearly or quite prevented, a conservative effect which we could scarcely expect to obtain if organic matter were the reducing agent. Meanwhile there is a gradual wasting away of the pyrites or positive pole, its sulphur being oxidised to sulphuric acid and its iron to sesquioxide of iron, or hematite, a substance very generally associated with gold nuggets. According to the original size of the pyritous mass, the protection it receives from the action of oxidising substances other than gold, the strength of the gold solution, length of exposure to it, the rate of supply and velocity of stream, will be the size of the gold nugget. As to the size of a pyritous mass necessary to produce in this manner a large nugget, it is by no means considerable. A mass of common pyrites (bisulphide of iron) weighing only 12 lbs. is competent for the construction of the famous 'Welcome Nugget,' an Australian find having weight equal to 152 lbs. avoirdupois. Such masses of pyrites are by no means uncommon in our drifts or the beds of our mountain streams. Thus we find that no straining of the imagination is required to conceive of this mode of formation for the huge masses of gold found in Australia in particular, such as the Welcome Nugget, 184 lbs. 9 oz.; the Welcome Stranger, a surface nugget, 190 lbs. after smelting; the Braidwood specimen nugget, 350 lbs., two-thirds gold; besides many other large masses of almost virgin gold which have been obtained from time to time in the alluvial diggings."

The author has made a number of experiments in the same direction, but more with the idea of demonstrating how possibly gold may in certain cases have been deposited in siliceous formations after such formations had solidified. Some of the results were remarkable and indeed unexpected. I found that I could produce artificial specimens of auriferous quartz from stone which had previously contained no gold whatever, also that it was not absolutely necessary that the stone so treated should contain any metallic sulphides.

The following was contributed by the author and is from the "Transactions" of the Australasian Institute of Mining Engineers for 1893: —

"THE DEPOSITION OF GOLD.

"The question as to how gold was originally deposited in our auriferous lodes is one to which a large amount of attention has been given, both by mineralogists and practical miners, and which has been hotly argued by those who held the igneous theory and those who pronounced for the aqueous theory. It was held by the former that as gold was not probably existent in nature in any but its metallic form, therefore it had been deposited in its siliceous matrix while in a molten state, and many ingenious arguments were adduced in support of this contention. Of late, however, most scientific men, and indeed many purely empirical inquirers (using the word empirical in its strict sense) have come to the conclusion that though the mode in which they were composed was not always identical, all lodes, including auriferous formations, were primarily derived from mineral-impregnated waters which deposited their contents in fissures caused either by the cooling of the earth's crust or by volcanic agency.

"The subject is one which has long had a special attraction for the writer, who has published several articles thereon, wherein it was contended that not only was gold deposited in the lodes from aqueous solution, but that some gold found in form of nuggets had not been derived from lodes but was nascent in its alluvial bed; and for this proof was afforded by the fact that certain nuggets have been unearthed having the shape of an adjacent pebble or angular fragment of stone indented in them. Moreover, no true nugget of any great size has ever been found in a lode such as the Welcome, 2159 oz., or the Welcome Stranger, 2280 oz.; while it was accidentally discovered some years ago that gold could be induced to deposit itself from its mineral salt to the metallic state on any suitable base, such as iron sulphide.

"Following out this fact, I have experimented with various salts of gold, and have obtained some very remarkable results. I have found it practicable to produce most natural looking specimens of auriferous quartz from stone which previously, as proved by assay, contained no gold whatever. Moreover, the gold, which penetrates the stone in a thorough manner, assumes some of the more natural forms. It is always more or less mammillary, but at times, owing to causes which I have not yet quite satisfied myself upon, is decidedly dendroidal, as may be seen in one of the specimens which I have submitted to members. Moreover, I find it possible to moderate the colour and to produce a specimen in which the gold shall be as ruddy yellow as in the ferro-oxide gangue of Mount Morgan, or to tone it to the pale primrose hue of the product of the Croydon mines.

"I note that the action of the bath in which the stone is treated has a particularly disintegrating effect on many of the specimens. Some, which before immersion were of a particularly flinty texture, became in a few weeks so friable that they could be broken up by the fingers. So far as my experiments have extended they have proved this, that it was not essential that the silica and gold should have been deposited at the one time in auriferous lodes. A non-auriferous siliceous solution may have filled a fissure, and, after solidifying, some volcanic disturbance may have forced water impregnated with a gold salt through the interstices of the lode formation, when, if the conditions were favourable, the gold would be deposited in metallic forms. I prefer, for reasons which will probably be understood, not to say exactly by what process my results are obtained, but submit specimens for examination.

"(1) Piece of previously non-gold bearing stone. Locality near Adelaide, now showing gold freely in mammillary and dendroidal form.

"(2) Stone from New South Wales, showing gold artificially introduced in interstices and on face.

"(3) Stone from West Australia, very glassy looking, now thoroughly impregnated with gold; the mammillary formation being particularly noticeable.

"(4) Somewhat laminated quartz from Victoria, containing a little antimony sulphide. In this specimen the gold not only shows on the surface but penetrates each of the laminations, as is proved by breaking.

"(5) Consists of fragments of crystallised carbonate of lime from Tarrawingee, in which the gold is deposited in spots, in appearance like ferrous oxide, until submitted to the magnifying glass.

"The whole subject is worthy of much more time than I can possibly give it. The importance lies in this: That having found how the much desired metal may have been deposited in its matrix, the knowledge should help to suggest how it may be economically extracted therefrom."

A very remarkable nugget weighing 16 3/4 oz. was sluiced from near the surface in one of my own mining properties at Woodside, South Australia, some years ago, which illustrated the nuclear theory very beautifully. This nugget is very irregular in shape, fretted and chased as though with a jeweller's graving tool, showing plainly the shape of the pyritous crystals on which it was formed while the interstices were filled with red hematite iron just as found in artificially formed nuggets on a sulphide of iron base. The author has a nugget from the same locality weighing about 1 1/2 oz. which exhibits in a marked degree the same characteristics, as indeed does most of the alluvial gold found in the Mount Lofty Ranges; also a nugget from near the centre of Australia weighing four ounces, in which the original crystals of pyrites are reproduced in gold just as an iron horse-shoe, placed in a launder through which cupriferously impregnated water flows, will in time be changed to nearly pure copper and yet retain its shape.

Now with regard to the four points I have put as to the apparent anomalies of occurrence of alluvial gold. The reason why alluvial gold is of finer quality as a rule than reef is probably because while gold and silver, which have a considerable affinity for each other, were presumably dissolved from their salts and held in solution in the same mineral water, they would in many cases not be deposited together, for the reason that silver is most readily deposited in the presence of alkalies, which would be found in excess in mineral waters coming direct from the basic rocks, while gold is induced to precipitate more quickly in acid solutions, which would be the character of the waters after they had been exposed to atmospheric action and to contact with organic matters.

This, then, may explain not only the comparatively greater purity of the alluvial gold, but also why big nuggets are found so far from auriferous reefs, and also why heavy masses of gold have been frequently unearthed from among the roots even of living trees, but more particularly in drifts containing organic matter, such as ancient timber.

All, then, that has been adduced goes to establish the belief that the birthplace of our gold is in certain of the earlier rocks comprising the earth's crust, and that its appearance as the metal we value so highly is the result of electro-chemical action, such as we can demonstrate in the laboratory.

CHAPTER VI

GOLD EXTRACTION

We now come to a highly important part of our subject, the practical treatment of ores and matrixes for the extraction of the metals contained. The methods employed are multitudinous, but may be divided into four classes, namely, washing, amalgamating with mercury, chlorinating, cyaniding and other leaching processes, and smelting. The first is used in alluvial gold and tin workings and in preparing some silver, copper, and other ores for smelting, and consists merely in separating the heavier metals and minerals from their gangues by their greater specific gravity in water. The second includes the trituration of the gangue and the extraction of its gold or silver by means of mercury. Chlorinating and leaching generally is a process whereby metals are first changed by chemical action into their mineral salts, as chloride of gold, nitrate of silver, sulphate of copper, and being dissolved in water are afterwards redeposited in the metallic form by means of well-known re-agents.

In really successful mining it is in the last degree important that the mode of extraction of metals in the most scientific manner should be thoroughly understood, but as a general rule the science of metallurgy is but very superficially grasped even by those whose special business it is to treat ore bodies in order to extract their metalliferous contents, and whether in quartz crushing mill, lixiviating, or smelting works there is much left to be desired in the method of treating our ores.

My attention was recently attracted to an article written by Mr. F. A. H. Rauft, M.E., from which I make the following extract:

He says, speaking of the German treatment of ores and the mode of procedure in Australia, "It is high time that Government stepped in and endeavoured by prompt and decisive action to bring the mining industry upon a sound and legitimate basis. Though our ranges abound in all kinds of minerals that might give employment to hundreds of thousands of people, mining is carried on in a desultory, haphazard fashion. There is no system, and the treatment of ores is of necessity handed over to the tender mercies of men who have not even an idea of what an intricate science metallurgy has become in older countries. During many years of practical experience I have never known a single instance where a lode, on being worked, gave a return according to assay, and I have never known any mine where some of the precious metals could not be found in the tailings or slag. The Germans employ hundreds of men in working for zinc which produces some two or three per cent to the ton; here the same percentage of tin could hardly be made payable, and this, mark you, is owing not to cheaper labour alone, but chiefly to the labour-saving appliances and the results of the researches of such gigantic intellects as Professor Kerl and many others, of whom we in this country never even hear. Go into any of the great mining works of central Germany, and you may see acres covered by machinery ingeniously constructed to clean, break, and sort, and ultimately deliver the ores into trucks or direct into the furnace, and the whole under the supervision of a youngster or two. When a parcel of ore arrives at any of the works, say Freiberg or Clausthal, it is carefully assayed by three or four different persons and then handed over to practical experts, who are expected to produce the full amount of previous metal according to assay; and if by any chance they do not, a fixed percentage of the loss is deducted from their salary; or, if the result is in excess of this assay which is more frequently the case, a small bonus is added to their pay. Compare this system with our own wasteful, reckless method of dealing with our precious metals, and we may hide our heads in very shame."

All really practical men will, I think, endorse Mr. Rauft's opinion. Well organised and conducted schools of mines will gradually ameliorate this unsatisfactory state of things, and I hope before long that we shall have none but qualified certificated men in our mines. In the meantime a few practical hints, particularly on that very difficult branch of the subject, the saving of gold, will, it is hoped, be found of service.

The extraction of gold from the soil is an industry so old that its first introduction is lost in the mist of ages. As before stated, gold is one of the most widely disseminated of the metals, and man, so soon as he had risen from the lowest forms of savagery, began to be attracted by the kingly metal, which he found to be easily fashioned into articles of ornament and use, and to be practically non-corrodable.

What we now term the dish or pan, then, doubtless generally a wooden bowl, was the appliance first used; but they had also an arrangement, somewhat like our modern blanket tables, over which the auriferous sand was passed by means of a stream of water. The sands of some of the rivers from which portions of the gold supply of the old world was derived are still washed over year after year in exactly the same manner as was employed, probably, thousands of years ago, the labour, very arduous, being often carried on by women, who, standing knee deep in water, pan off the sand in wooden bowls much as the digger in modern alluvial fields does with his tin dish. The resulting gold often consists of but a grain or two of fine dust-gold, which is carefully collected in quills, and so exported or traded for goods.

The digger of to-day having discovered payable alluvial dirt at such a depth as to permit of its being profitably worked by small parties of men with limited or no capital, procures first a half hogshead for a puddling tub, a "cradle," or "long tom," and tin dish. The "wash dirt," as the auriferous drift is usually termed, contains a considerable admixture of clay of a more or less tenacious character, and the bulk of this has to be puddled and so disintegrated before the actual separation of the gold is attempted in the cradle or dish. This is done in the tub by constantly stirring with a shovel, and changing the water as it becomes charged with the floating argillaceous, or clayey, particles. The gravel is then placed in the hopper of the cradle which separates the larger stones and pebbles, the remainder passing down over inclined ledges as the cradle is slowly rocked and supplied with water. At the bottom of each ledge is a riffle to arrest the particles of gold. Sometimes, when the gold is very fine, amalgamated copper plates are introduced and the lower ledges are covered with green baize to act as blanket tables and catch gold which might otherwise be lost.

A long tom is a trough some 12 feet in length by 20 inches in width at the upper end, widening to 30 inches at the lower end; it is about 9 inches deep and has a fall of 1 inch to a foot. An iron screen is placed at the lower end where large stones are caught, and below this screen is the riffle box, 12 feet long, 3 feet wide, and having the same inclination as the upper trough. It is fitted with several riffles in which mercury is sometimes placed.

Much more work can be done with this appliance than with the cradle, which it superseded. Of course, the gold must be coarse and water plentiful.

When, however, the claim is paying, and the diggings show signs of some permanency, a puddling machine is constructed. This is described in the chapter called "Rules of Thumb."

Hydraulicing and ground sluicing is a very cheap and effective method of treating large quantities of auriferous drift, and, given favourable circumstances, such as a plentiful supply of water with good fall and extensive loose auriferous deposits, a very few grains to the ton or load can be made to give payable returns. The water is conveyed in flumes, or pipes to a point near where it is required, thence in wrought iron pipes gradually reduced in size and ending in a great nozzle somewhat like that of a fireman's hose. The "Monitor," as it is sometimes called, is generally fixed on a movable stand, so arranged that the strong jet of water can be directed to any point by a simple adjustment. A "face" is formed in the drift, and the water played against the lower portion of the ledge, which is quickly undermined, and falls only to be washed away in the stream of water, which is conducted through sluices with riffles, and sometimes over considerable lengths of amalgamated copper plates. This class of mining has been most extensively carried out in California and New Zealand, and some districts of Victoria, but the truly enormous drifts of the Shoalhaven district in New South Wales must in the near future add largely to the world's gold supply. These drifts which are auriferous from grass roots to bed rock extend for nearly fifty miles, and are in places over 200 feet deep. Want of capital and want of knowledge has hitherto prevented their being profitably worked on a large scale.

The extraction of reef gold from its matrix is a much more complicated process, and the problem how most effectively to obtain that great desideratum—a complete separating and saving operation—is one which taxes the skill and evokes the ingenuity of scientific men all over the world. The difficulty is that as scarcely any two gangues, or matrixes, are exactly alike, the treatment which is found most effective on one mine will often not answer in another. Much also depends on the proportion of gold to the ton of rock under treatment, as the most scientific and perfect processes of lixiviation hitherto adopted will not pay, even when all other conditions are favourable, if the amount of gold is much under half an ounce to the ton and even then will leave but a very small profit. If, however, the gold is "free," and the lode large, a very few pennyweights (or "dollars," as the Americans say) to the ton will pay handsomely. The mode of extraction longest in vogue, and after all the cheapest and most effective, for free milling ores where the gold is not too fine, is amalgamation with mercury, which metal has a strong affinity for gold, silver, and copper.

As to crushing appliances, I shall not say much. "Their name is legion for they are many," and the same may be said of concentrators. It may be old-fashioned, but I admit my predilection is still in favour of the stamper-battery, for the reason that though it may be slower in proportion to the power employed, it is simple and not liable to get out of order, a great advantage when one has so often to depend on men who bring to their work a supply principally of main strength and stupidity. For the same reason I prefer the old draw and lift, and plunger pumps to newer but more complicated water-lifters.

On both these points, however, I am constrained to admit that my opinion has recently been somewhat shaken.

I have lately seen two appliances which appear to mark a new era in the scientific progress of mining. One is the "Griffin Mill," the other the "Lemichel Siphon Elevateur."

The first is in some respects on the principle of the Huntingdon Mill. The latter, if the inventor may be believed and the results seem to show he can be, will be a wonderful factor in developing not only mining properties where a preponderance of water is the trouble, but also in providing an automatic, and therefore extremely cheap, mode of water-raising and supply, which in simplicity is thus far unexampled. Atmospheric pressure alone is relied on. The well-known process of the syphon is the basis, but with this essential difference, that a large proportion of the water drawn up to the apex of the syphon is super-elevated to heights regulated by the fall obtained in the outlet leg. This elevation can be repeated almost indefinitely by returning the waste water to the reservoirs.

The Lemichel Syphon is a wonderful, yet most simple application of natural force. The inlet leg of the syphon is larger in diameter than the outlet leg, and is provided at the bottom with a valve or "clack." The outlet leg has a tap at its base. At the apex are two chambers, with an intermediary valve, regulated by a counterpoise weighted lever. The first chamber has also a vertical valve and pipe.

When the tap of the outlet leg is turned, the water flows as in an ordinary syphon, but owing to the rapid automatic opening and shutting of the valve in the first chamber about 45 per cent of the water is diverted, and may be raised to a height of many feet above the top of the syphon.

It need not be impressed on practical men that if this invention will perform anything like what is claimed for it, its value can hardly be calculated. After a careful inspection of the appliance in operation, I believe it will do all that is stated.

Another invention is combined with this which, by a very small expenditure of fuel, will enable the first point of atmospheric pressure to be attained. In this way the unwatering of mines may be very inexpensively effected, or water for irrigation purposes may be raised from an almost level stream.

The Griffin Mill is a centrifugal motion crusher with one roller only, which, by an ingenious application of motive force, revolves in an opposite direction to its initial momentum, and which evolves a force of 6000 lb. against the tire, which is only 30 inches in diameter. For hard quartz the size should be increased by at least 6 inches. It is claimed for this mill that it will pulverise to a gauge of 900 holes to the square inch from 1 1/2 to 2 1/2 tons per hour, or, say roughly, 150 tons per week.

The Huntingdon mill is a good crusher and amalgamator where the material to be operated on is comparatively soft, but does not do such good work when the stone is of a hard flinty nature.

A No. 4 Dodge stone-breaker working about 8 hours will keep a five-foot Huntingdon mill going 24 hours, and an automatic feeder is essential. For that matter both are almost essential for an ordinary stamper battery, and will certainly increase the crushing capacity and do better work from the greater regularity of the feed.

A 10 h.-p. (nominal) engine of good type is sufficient for Huntingdon mill, rock breaker, self-feeder and steam pump. A five-foot mill under favourable circumstances will crush about as much as eight head of medium weight stamps.

The Grusonwek Ball Mills, made by Krupp of Germany, also that made by the Austral Otis Company, Melbourne, are fast and excellent crushing triturating appliances for either wet or dry working, but are specially suited only for ores when the gold is fine and evenly distributed in the stone. The trituration is effected by revolving the stone in a large cylinder together with a number of steel balls of various sizes, the attrition of which with the rock quickly grinds it to powder of any required degree of fineness.

More mines have been ruined by bad mill management probably than by bad mining, though every experienced man must have seen in his time many most flagrant instances of bungling in the latter respect. Shafts are often sunk on the wrong side of the lode or too near or too far away therefrom, while instances have not been wanting where the (mis) manager has, after sinking his shaft, driven in the opposite direction to that where the lode should be found.

A common error is that of erecting machinery before there is sufficient ore in sight to make it certain that enough can be provided to keep the plant going. In mines at a distance from the centre of direction it is almost impossible to check mistakes of this description, caused by the ignorance or over sanguineness of the mine superintendent, and they are often as disastrous as they are indefensible. Another fertile source of failure is the craze for experimenting with untried inventions, alleged to be improvements on well-known methods.

A rule in the most scientific of card games, whist, is "when in doubt lead trumps." It might be paraphrased for mining thus: "When in doubt about machinery use that which has been proved." Let some one else do the experimenting.

The success of a quartz mine depends as much on favourable working conditions as on its richness in gold. Thus it may be that a mine carrying 5 or 6 oz. of gold to the ton but badly circumstanced as to distance, mountainous roads, lack of wood and water, in some cases a plethora of the latter, or irregularly faulted country, may be less profitable than another showing only 5 or 6 dwt., but favourably situated.

It is usually desirable to choose for the battery site, when possible, the slope of a hill which consists of rock that will give a good foundation for your battery.

The economical working depends greatly on the situation, which is generally fixed more or less, in the proximity of the water. The advantages of having ample water for battery purposes, or of using water as a motive power, are so great that it is very often desirable to construct a tramway of considerable length, when, by so doing, that power can be utilised; hence most quartz mills are placed near streams, or in valleys where catchment dams can be effectively constructed, except, of course, in districts where much water has to be pumped from the mine.

If water-power can be used, the water-motor will necessarily be placed as low as possible, in order to obtain the fullest available power. One point is essential. Special care must be taken to keep the appliances above the flood-level. If the water in the stream is not sufficient to carry off the tailings, the battery should be placed at such a height as to leave sufficient slope for tailings' dumps. This is more important when treating ore of such value that the tailings are worth saving for secondary treatment. In this case provision should be made for tailings, dams, or slime pits.

Whether the battery is worked by water, steam, or gas power, an ample supply of water is absolutely necessary, at least until some thoroughly effective mode of dry treatment is established. If it can be possibly arranged the water should be brought in by gravitation, and first cost is often least cost; but where this is impossible, pumps of sufficient capacity, not only to provide the absolute quantity used, but to meet any emergency, should be erected.

The purer the water the better it will be for amalgamating purposes, and in cold climates it is desirable to make provision for heating the water supplied to the battery. This can be done by means of steam from the boiler led through the feed tanks; but where the boiler power is not more than required, waste steam from the engine may be employed, but care must be taken that no greasy matter comes in contact with the plates. The exhaust steam from the engine may be utilised by carrying it through tubes fitted in an ordinary 400 gallon tank.

Reducing appliances have often to be placed in districts where the water supply is insufficient for the battery. When this is so every available means must be adopted for saving the precious liquid, such as condensing the exhaust steam from the engine. This may be done by conducting it through a considerable length of ordinary zinc piping, such as is used for carrying the water from house roofs. Also tailings pits should be made, in which the tailings and slimes are allowed to settle, and the cleared water is pumped back to be again used. These pits should, where practicable, be cemented. It is usual, also, to have one or two tailings dams at different levels; the tailings are run into the upper dam, and are allowed to settle; the slimes overflow from it into the lower dam, and are there deposited, while the cleared water is pumped back to the battery. Arrangements are made by which all these reservoirs can be sluiced out when they are filled with accumulated tailings. It is well not to leave the sluicing for too long a period, as when the slimes and tailings are set hard they are difficult to remove.

Where a permanent reducing plant is to be erected, whatever form of mill may be adopted, it is better for many reasons to use automatic ore feeders. Of these the best two I have met are the "Tulloch" and "Challenge" either of which can be adapted to any mill and both do good work.

By their use the reducing capacity of the mill is increased, and the feeding being regular the wear and tear is decreased, while by the regulated feeding of the "pulp" in the battery box or mortar can be maintained at any degree of consistency which may be found desirable, and thus the process of amalgamation will be greatly facilitated. The only objection which can be urged against the automatic feeder is that the steel points of picks, gads, drills, and other tools may be allowed to pass into the mortar or mill, and thus cause considerable wear and tear. This, I think, can be avoided by the adoption of the magnet device, described in "Rules of Thumb."

There are many mines where 3 to 4 dwt. of gold cover all the cost, the excess being clear profit. In fact there are mines which with a yield of 1 1/2 to 2 dwt. a ton, and crushing with water power, have actually yielded large profits. On the other hand, mines which have given extraordinary trial crushings have not paid working expenses. Everything depends on favourable local conditions and proper management.

Having decided what class of crushing machinery you will adopt, the first point is to fix on the best possible site for its erection. This requires much judgment, as success or failure may largely depend on the position of your machinery. One good rule is to get your crusher as reasonably high as possible, as it is cheaper to pump your feed water a few feet higher so as to get a good clear run for your tailings, and also to give you room to erect secondary treatment appliances, such as concentrators and amalgamators below your copper plates and blanket strakes.

Next, and this is most important, see that your foundations are solid and strong. A very large number of the failures of quartz milling plants is due to neglect of this rule.

I once knew a genius who erected a 10-Lead mill in a new district, and who adopted the novel idea of placing a "bed log" laterally beneath his stampers. The log was laid in a little cement bed which, when the battery started, was not quite dry. The effect was comical to every one but the unfortunate owners. It was certainly the liveliest, but at the same time one of the most ineffective batteries I have seen.

In a stamp mill the foundations are usually made of hard wood logs about 5 to 6 feet long, set on end, the bottom end resting on rock and set round with cement concrete. These are bolted together, and the "box" or mortar is bolted to them. The horizontal logs to carry the "horses" or supports for the battery frame should also be of good size, and solidly and securely bolted. The same applies to your engine-bed, but whether it be of timber, or mason work, above all things provide that the whole of your work is set out square and true to save after-wear and friction.

Considerable difference of opinion exists as to the most effective weight for stamps. My experience has been that this largely depends on the nature of your rock, as does also the height for the drop. I have usually found that with medium stamps, say 7 to 7 1/2 cwt. with fair drop and lively action, about 80 falls per minute, the best results were obtained, but the tendency of modern mill men is towards the heavier stamps, 9 cwt. and even heavier.

To find the horse-power required to drive a battery, multiply the weight of one stamp by the number of stamps in the battery; the height of lift in feet by the number of lifts per minute; add one-third of the product for friction, and the result will be the number of feet-lbs. per minute; divide this by 33,000 which is the number of feet-lbs. per minute equal to 1 h.-p. and the result will be the h.-p. required. Thus if a stamp weighs 800 lb. and you have five in the box, and each stamp has a lift of 9 in. = 0.75 ft. and strikes 80 blows per minute, then 800 x 5 x 0.75 x 80 = 240,000; one-third of 240,000 = 80,000 which added to 240,000 = 320,000; and 320,000 divided by 33,000 = 9.7 h.-p. or 1.9 h.-p. each stamp.

The total weight of a battery, including stamper box, stampers, etc., may be roughly estimated at about 1 ton per stamp. Medium weight stampers, including shank cam, disc, head, and shoe, weigh from 600 to 700 lb., and need about 3/4 h.-p. to work them.

The quantity of water required for the effective treatment of gold-bearing rock in a stamper battery varies according to the composition of the material to be operated upon, but generally it is more than the inexperienced believe. For instance, "mullocky" lode stuff, containing much clayey matter or material carrying a large percentage of heavy metal, such as titanic iron or metallic sulphides, will need a larger quantity of water per stamp than clean quartz. A fair average quantity would be 750 to 1000 gallons per hour for each box of five stamps. In general practice I have seldom found 1000 gallons per hour more than sufficient.

As to the most effective mesh for the screen or grating no definite rule can be given, as that depends so largely on the size of the gold particles contained in the gangue. The finer the particles the closer must be the mesh, and nothing but careful experiment will enable the battery manager to decide this most important point. The American slotted screens are best; they wear better than the punched gratings and can be used of finer gauge. Woven steel wire gauze is employed with good effect in some mills where especially fine trituration is required. This class of screen requires special care as it is somewhat fragile, but with intelligent treatment does good work.

The fall or inclination of the tables, both copper and blanket strakes, is also regulated by the class of ore. If it should be heavy then the fall must be steeper. A fair average drop is 3/4 inch to the foot. Be careful that your copper tables are thoroughly water-tight, for remember you are dealing with a very volatile metal, quicksilver; and where water will percolate mercury will penetrate.

The blanket tables are simply a continuation of the mercury tables, but covered with strips of coarse blanket, green baize, or other flocculent material, intended to arrest the heavier metallic particles which, owing to their refractory nature, have not been amalgamated.

The blanket table is, however, a very unsatisfactory concentrator at best, and is giving place to mechanical concentrators of various descriptions.

An ancient Egyptian gold washing table was used by the Egyptians in treating the gold ores of Lower Egypt. The ore was first ground, it is likely by means of some description of stone arrasts and then passed over the sloping table with water, the gold being retained in the riffles. In these the material would probably be mechanically agitated. Although for its era ingenious it will be plain to practical men that if the gold were fine the process would be very ineffective. Possibly, but of this I have no evidence, mercury was used to retain the gold on the riffles, as previously stated. This method of saving the precious metal was known to the ancients.

At a mine of which I was managing director the lode was almost entirely composed of sulphide of iron, carbonate of lime or calcspar, with a little silica. In this case it has been found best to crush without mercury, then run the pulp into pans, where it is concentrated. The concentrates are calcined in a common reverberatory furnace, and afterwards amalgamated with mercury in a special pan, the results as to the proportion of gold extracted being very satisfactory; but it does not therefore follow that this process would be the most suitable in another mine where the lode stuff, though in some respects similar, yet had points of difference.

I was lately consulted with respect to the treatment of a pyritic ore in a very promising mine, but could not recommend the above treatment, because though the pyrites in the gangue was similar, the bulk of the lode consisted of silica, consequently there would be a great waste of power in triturating the whole of the stuff to what, with regard to much of it, would be an unnecessary degree of fineness. I am of opinion that in cases such as this, where it is not intended to adopt the chlorination or cyanogen process, it will be found most economical to crush to a coarse gauge, concentrate, calcine the concentrates, and finally amalgamate in some suitable amalgamator.

Probably for this mode of treatment Krom rolls would be found more effective reducing agents than stampers, as with them the bulk of the ore can be broken to any required gauge and there would consequently be less loss in "slimes."

The great art in effective battery work is to crush your stuff to the required fineness only, and then to provide that each particle is brought into contact with the mercury either in box, trough, plate, or pan. To do this the flow of water must be carefully regulated; neither so much must be used as to carry the stuff off too quickly nor so little as to cause the troughs and plates to choke. In cold weather the water may be warmed by passing the feed-pipe through a tank into which the steam from the engine exhausts, and this will be found to keep the mercury bright and lively. But be careful no engine oil or grease mingles with the water, as grease on the copper tables will absolutely prevent amalgamation.

The first point, then, is to crush the gangue effectively, the degree of fineness being regulated by the fineness of the gold itself. This being done, then comes the question of saving the gold. If the quartz be clean, and the gold unmixed with base metal, the difficulty is small. All that is required is to ensure that each particle of the Royal metal shall be brought into contact with the mercury. The main object is to arrest the gold at the earliest possible stage; therefore, if you are treating clean stone containing free gold, either coarse or fine, I advise the use of mercury in the boxes, for the reason that a considerable proportion of the gold will be caught thereby, and settling to the bottom, or adhering to amalgamated plates in the boxes, where such are used, will not be afterwards affected by the crushing action, which might otherwise break up, or "flour," the mercury. On the whole, I rather favour the use of mercury in the box at any time, unless the ore is very refractory—that is, contains too great a proportion of base metals, particularly sulphides of iron, arsenic, etc., when the result will not be satisfactory, but may entail great loss by the escape of floured mercury carrying with it particles of gold. Here only educated intelligence, with experience, will assist the battery manager to adopt the right system.

The crushed stuff—generally termed the "pulp"—passes from the boxes through the "screens" or "gratings," and so on to the "tables"— i.e., sheets of copper amalgamated on the upper surface with mercury, and sometimes electroplated with silver and afterwards treated with mercury. Unless the quartz is very clean, and, consequently light, I am opposed to the form of stamper box with mercury troughs cast in the "lip," nor do I think that a trough under the lip is a good arrangement, as it usually gets so choked and covered with the heavy clinging base metals as to make it almost impossible for the gold to come in contact with the mercury. It will be found better where the gold is fine, or the gangue contains much base metal, to run the pulp from the lip of the battery into a "distributor."

The distributor is a wooden box the full width of the "mortar," having a perforated iron bottom set some three to four inches above the first copper plate, which should come up under the lip. The effect of this arrangement is that the pulp is dashed on the plate by the falling water, and the gold at once coming in contact with the mercury begins to accumulate and attract that which follows, till the amalgam becomes piled in little crater-shaped mounds, and thus 75 per cent of the gold is saved on the top plate.

I have tried a further adaptation of this process when treating ores containing a large percentage of iron oxide, where the bulk of the gold is impalpably fine, and contained in the "gossan." At the end of the blanket table, or at any point where the crushed stuff last passes before going to the "tailings heap," or "sludge pit," a "saver" is placed. The saver is a strong box about 15 in. square by 3 ft. high, one side of which is removable, but must fit tight. Nine slots are cut inside at 4 in. apart, and into these are fitted nine square perforated copper plates, having about eighty to a hundred 1/4 in. holes in each; the perforations should not come opposite each other. These plates are to be amalgamated on both sides with mercury, in which a very little sodium has been placed (if acid ores are being treated, zinc should be employed in place of sodium, and to prevent the plates becoming bare, if the stuff is very poor, thick zinc amalgam may be used with good effect; but in that case discontinue the sodium, and occasionally, if required, say once or twice in the day, mix an ounce of sulphuric acid in a quart of water and slowly pour it into the launder above the saver). Underneath the "saver" you require a few riffles, or troughs, to catch any waste mercury, but if not overfed there should be no waste. This simple appliance, which is automatic and requires little attention, will sometimes arrest a considerable quantity of gold.

We now come to the subsidiary processes of battery work, the "cleaning" of plates, and "scaling" same when it is desired to get all the gold off them, the cleaning and retorting of amalgam, and of the mercury, smelting gold, etc.

Plates should be tenderly treated, kept as smooth as possible, and when cleaning up after crushing, in your own battery, the amalgam—except, say, at half-yearly intervals—should be removed with a rubber only; the rubber is simply a square of black indiarubber or soft pine wood.

When crushing rich ore, and you want to get nearly all the gold off your plates, the scraper may be resorted to. This is usually made by the mine blacksmith from an old flat file which is cut in half, the top turned over, beaten out to a sharp blade, and kept sharp by touching it up on the grinding-stone. This, if carefully used, will remove the bulk of the amalgam without injury to the plate.

Various methods of "scaling" plates will be found among "Rules of Thumb."

Where base metals are present in the lode stuff frequent retortings of the mercury, say not less than once a month, will be found to have a good effect in keeping it pure and active. For this purpose, and in order to prevent stoppage of the machinery, a double quantity is necessary, so that half may be used alternately. Less care is required in retorting the mercury than in treating the amalgam, as the object in the one case is more to cleanse the metal of impurities than to save gold, which will for the most part have been extracted by squeezing through the chamois leather or calico. A good strong heat may therefore at once be applied to the retort and continued, the effect being to oxidise the arsenic, antimony, lead, etc., which, in the form of oxides, will not again amalgamate with the mercury, but will either lie on its surface under the water, into which the nozzle of the retort is inserted, or will float away on the surface of the water. I have also found that covering the top of the mercury with a few inches of broken charcoal when retorting has an excellent purifying effect.

In retorting amalgam, much care and attention is required.

First, never fill the retort too full, give plenty of room for expansion; for, when the heat is applied, the amalgam will rise like dough in an oven, and may be forced into the discharge pipe, the consequence being a loss of amalgam or the possible bursting of the retort. Next, be careful in applying the heat, which should be done gradually, commencing at the top. This is essential to prevent waste and to turn out a good-looking cake of gold, which all battery managers like to do, even if they purpose smelting into bars.

Sometimes special difficulties crop up in the process of separating the gold from the amalgam. At the first "cleaning up" on the Frasers Mine at Southern Cross, West Australia, great consternation was excited by the appearance of the retorted gold, which, as an old miner graphically put it, was "as black as the hind leg of a crow," and utterly unfit for smelting, owing to the presence of base metals. Some time after this I was largely interested in the Blackborne mine in the same district when a similar trouble arose. This I succeeded in surmounting, but a still more serious one was too much for me— i.e., the absence of payable gold in the stone. I give here an extract from the *Australian Mining Standard*, of December 9th, 1893, with reference to the mode of cleaning the amalgam which I adopted.

NEW METHOD OF SEPARATING GOLD FROM IMPURE AMALGAM.

I had submitted to me lately a sample of amalgam from a mine in West Australia which amalgam had proved a complete puzzle to the manager and amalgamator. The Mint returns showed a very large proportion of impurity, even in the smelted gold. When retorted only, the Mint authorities refused to take it after they had treated two cakes, one of 119 oz., which yielded only 35 oz. 5 dwt. standard gold, and one of 140 oz., which gave 41 oz. 10 dwt. The gold smelted on the mine was nearly as bad proportionately. Thus, 128 oz. smelted down at the Mint to 87 oz. 8 dwt. and 109 oz. to 55 oz. 10 dwt. The impurity was principally iron, a most unusual thing in my experience, and was due to two causes revealed by assay of the ore and analysis of the mine water, viz., an excess of arsenate of iron in the stone, and the presence in large proportions of mineral salts, principally chloride of Calcium CaCl., sodium NaCl, and magnesium $MgCl_2$, in the mine water used in the battery. The exact analysis of the water was as follows: —

> Carbonate of Iron $FeCO_3$ 2.76 grains per gallon Carbonate of Calcium $CaCO_3$ 7.61 grains per gallon Sulphate of Calcium $CaSO_4$ 81.71 grains per gallon Chloride of Calcium $CaCl_2$ 2797.84 grains per gallon Chloride of Magnesium $MgCl_2$ 610.13 grains per gallon Chloride of Sodium or Common Salt NaCl 5072.65 grains per gallon Total solid matter 8572.70 = 19.5 oz. to the gallon.

It will be seen, then, that this water is nearly four times more salt that that of the sea. The effect of using a water of this character, as I have previously found, is to cause the amalgamation of considerable quantities of iron with the gold as in this case.

I received 10 oz. of amalgam, and having found what constituted its impurities proceeded to experiment as to its treatment. When retorted on the mine it was turned out in a black cake so impure as almost to make it impossible to smelt properly. I found the same result on first retorting, and after a number of experiments which need not be recapitulated though some were fairly effective, I hit on the following method, which was found to be most successful and will probably be so found in other localities where similarly unfavourable conditions prevail.

I took a small ball of amalgam, placed it in a double fold of new fine grained calico, and after soaking in hot water put it under a powerful press. The weight of the ball before pressing was 1583 gr. From this 383 gr. of mercury was expressed and five-eighths of a grain of gold was retorted from this expressed mercury. The residue, in the form of a dark, grey, and very friable cake, was powdered up between the fingers and retorted, when it became a brown powder; it was afterwards calcined on a flat sheet in the open air; result, 510 gr. of russet-coloured powder. Smelted with borax, the iron oxide readily separated with the slag; result, 311 gr. gold 871-1000 fine; a second smelting brought this up to 914-1000 fine. Proportion of smelted gold to amalgam, one-fifth.

The principal point about this mode of treatment is the squeezing out of the mercury, whereby the amalgam goes into the retort in the form of powder, thus preventing the slagging of the iron and enclosure of the gold. The second point of importance is thorough calcining before smelting.

Of course it would be practicable, if desired, to treat the powder with hydrochloric acid, and thus remove all the iron, but in a large way this would be too expensive, and my laboratory treatment, though necessarily on a small scale, was intended to be on a practical basis.

The amalgam at this mine was in this way afterwards treated with great success.

For the information of readers who do not understand the chemical symbols it may be said that

> $FeCO_3$ is carbonate of iron; $CaCO_3$ is carbonate of calcium; $CaSO_4$ is sulphate of calcium; $CaCl_2$ is chloride of calcium; $MgCl_2$ is chloride of magnesium; $NaCl$ is chloride of sodium, or common salt.

CHAPTER VII

GOLD EXTRACTION – SECONDARY PROCESSES AND LIXIVIATION

Before any plan is adopted for treating the ore in a new mine the management should very seriously and carefully consider the whole circumstances of the case, taking into account the quantity and quality of the lode stuff to be operated on, and ascertain by analysis what are its component parts, for, as before stated, the treatment which will yield most satisfactory results with a certain class of gangue on one mine will sometimes, even when the material is apparently similar, prove a disastrous failure in another. Some time since I was glad to note that the manager of a prominent mine strongly discountenanced the purchase of any extracting plant until he was fully satisfied as to the character of the bulk of the ore he would have to treat. It would be well for the pockets of shareholders and the reputation of managers, if more of our mine superintendents followed this prudent and sensible course.

Having treated on gold extraction with mercury by amalgamated plates and their accessories, something must be said about secondary modes of saving in connection with the amalgamation process. The operations described hitherto have been the disintegration of the gold-bearing material and the extraction therefrom of the coarser free gold. But it must be understood that most auriferous lode stuff contains a proportion of sulphides of various metals, wherein a part of the gold, usually in a very finely divided state, is enclosed, and on this gold the mercury has no influence. Also many lodes contain hard heavy ferric ores, such as titanic iron, tungstate of iron, and hematite, in which gold is held. In others, again, are found considerable quantities of soft powdery iron oxide or "gossan," and compounds such as limonite, aluminous clay, etc., which, under the action of the crushing mill become finely divided and float off in water as "slimes," carrying with them atoms of gold, often microscopically small. To save the gold in such matrixes as these is an operation which even the best of our mechanical appliances have not yet fully accomplished.

Where there is not too great a proportion of base metals on which the solvent will act, and when the material is rich enough in gold to pay for the extra cost of treatment, chlorination or cyanisation are the best modes of extraction yet practically adopted.

Presuming, however, that we are working by the amalgamation process, and have crushed our stone and obtained the free gold, the next requirement is an effective concentrator. Of these there are many before the public, and some do excellent work, but do not act equally well in all circumstances. The first and most primitive is the blanket table, previously mentioned; but it can hardly be said to be very effective, and requires constant attention and frequent changing and washing of the strips of blanket.

Instead of blanket tables percussion tables are sometimes used, to which a jerking motion is given against the flow of the water and pulp, and by this means the heavier minerals are gathered towards the upper part of the table, and are from thence removed from time to time as they become concentrated.

I have seen this appliance doing fairly good work, but it is by no means a perfect concentrator.

Another form of "shaking table" is one in which the motion is given sideways, and this, whether amalgamated, or provided with small riffles, or covered with blanket, keeps the pulp lively and encourages the retention of the heavier particles, whether of gold or base metals containing gold. There has also been devised a rocking table the action of which is analogous to that of the ordinary miner's cradle. This appliance, working somewhat slowly, swings on rockers from side to side, and is usually employed in mills where, owing to the complexity of the ore, difficulties have been met with in amalgamating the gold. Riffles are provided and even very fine gold is sometimes effectively recovered by their aid.

The Frue vanner will, as a rule, act well when the pulp is sufficiently fine. It is really a adaptation of an old and simple apparatus used in China and India for washing gold dust from the sands of rivers. The original consisted of an endless band of strong cloth or closely woven matting, run on two horizontal rollers placed about seven feet apart, one being some inches lower than the other. The upper is caused to revolve by means of a handle. The cloth is thus dragged upwards against a small stream of water and sand fed to it by a second man, the first man not only turning the handle but giving a lateral motion to the band by means of a rope tied to one side.

Chinamen were working these forerunners of the Frue vanner forty years ago in Australia, and getting fair returns.

The Frue vanner is an endless indiarubber band drawn over an inclined table, to which a revolving and side motion is given by ingenious automatic mechanism, the pulp being automatically fed from the upper end, and the concentrates collected in a trough containing water in which the band is immersed in its passage under the table; the lighter particles wash over the lower end. The only faults with the vanner are—first, it is rather slow; and secondly, though so ingenious it is just a little complicated in construction for the average non-scientific operative.

Of pan concentrators there is an enormous selection, the principle in most being similar—i.e., a revolving muller, which triturates the sand, so freeing the tiny golden particles and admitting of their contact with the mercury. The mistake with respect to most of these machines is the attempt to grind and amalgamate in one operation. Even when the stone under treatment contains no deleterious compounds the simple action of grinding the hard siliceous particles has a bad effect on the quicksilver, causing it to separate into small globules, which either oxidising or becoming coated with the impurities contained in the ore will not reunite, but wash away in the slimes and take with them a percentage of the gold. As a grinder and concentrator, and in some cases as an amalgamator, when used exclusively for either purpose, the Watson and Denny pan is effective; but although successfully used at one mine I know, the mode there adopted would, for reasons previously given, be very wasteful in many other mines.

There is considerable misconception, even among men with some practical knowledge, as to the proper function of these secondary saving appliances; and sometimes good machines are condemned because they will not perform work for which they were never intended. It cannot be too clearly realized that the correct order of procedure for extracting the gold held in combination with base metals is—first, reduction of the particles to a uniform gauge and careful concentration only; next, the dissipation, usually by simple calcination, of substances in the concentrates inimical to the thorough absorption of the gold by the mercury; and lastly, the amalgamation of the gold and mercury.

For general purposes, where the gangue has not been crushed too fine, I think the Duncan pan will usually be found effective in saving the concentrates. In theory it is an enlargement of the alluvial miner's tin dish, and the motion imparted to it is similar to the eccentric motion of that simple separator.

The calcining may be effectively carried out in an ordinary reverberatory furnace, the only skill required being to prevent over roasting and so slagging the concentrates; or not sufficiently calcining so as to remove all deleterious constituents; the subject, however, is fully treated in Chapter VIII.

For amalgamating I prefer some form of settler to any further grinding appliance, but I note also improvements in the rotary amalgamating barrel, which, though slow, is, under favourable conditions, an effective amalgamator. The introduction of steam under pressure into an iron cylinder containing a charge of concentrates with mercury is said to have produced good results, and I am quite prepared to believe such would be the case, as we have long known that the application of steam to ores in course of amalgamation facilitates the process considerably.

Some seventeen years since I was engaged on the construction of a dry amalgamator in which sublimated mercury was passed from a retort through the descending gangue in a vertical cylinder, the material thence falling through an aperture into a revolving settler, the object being to save water on mines in dry country. The model, about quarter size, was completed when my attention was called to an American invention, in which the same result was stated to be attained more effectively by blowing the mercury spray through the triturated material by means of a steam jet. I had already encountered a difficulty, since found so obstructive by experimentalists in the same direction, that is, the getting of the mercury back into its liquid metallic form. This difficulty I am now convinced can be largely obviated by my own device of using a very weak solution of sulphuric acid (it can hardly be too weak) and adding a small quantity of zinc to the mercury. It is perfectly marvellous how some samples of mercury "sickened" or "floured" by bad treatment, may be brought back to the bright limpid metal by a judicious use of these inexpensive materials.

Thus it will probably be found practicable to crush dry and amalgamate semi-dry by passing the material in the form of a thin pasty mass to a settler, as in the old South American arrastra, and, by slowly stirring, recover the mercury, and with it the bulk of the gold.

The following is from the *Australian Mining Standard*, and was headed "Amalgamation Without Overflow":

"Recent experiments at the Ballarat School of Mines have proved that a deliverance from difficulties is at hand from an unexpected quarter. The despised Chilian mill and Wheeler pan, discarded at many mines, will solve the problem, but the keynote of success is amalgamation without overflow. Dispense with the overflow and the gold is saved.

"Two typical mines—the Great Mercury Proprietary Gold Mine, of Kuaotunu, N.Z., the other, the Pambula, N.S.W.—have lately been conducting a series of experiments with the object of saving their fine gold in an economical manner. The last and best trials made by these companies were at the Ballarat School of Mines, where amalgamation without overflow was put to a crucial test, in each case with the gratifying result that ninety-six per cent of the precious metal was secured. What this means to the Great Mercury Mine, for instance, can easily be imagined when it is understood that notwithstanding all the latest gold-saving adjuncts during the last six months 1260 tons of ore, worth 4l. 17s. 10 23d. *a ton, have been put through for a saving of* 1l. 9s. 1 23d. only; or in other words over two-thirds of the gold has gone to waste (for the time being) in the tailings, and in the tailings at the present moment lie the dividends that should have cheered shareholders' hearts.

"And now for the *modus operandi*, which, it must be remembered, is not hedged in by big royalties to any one, rights, patent or otherwise. The ore to be treated is first calcined, then put through a rock-breaker or stamper battery in a perfectly dry state. If the battery is used, ordinary precautions, of course, must be taken to prevent waste, or the dust becoming obnoxious to the workmen. The ore is then transferred to the Chilian mill and made to the consistency of porridge, the quicksilver being added. When the principal work of amalgamation is done (experience soon teaching the amount of grinding necessary), from the Chilian mill the paste (so to say) is passed to a Wheeler or any other good pan of a similar type, when the gold-saving operation is completed."

This being an experiment in the same direction as my own, I tried it on a small scale. I calcined some very troublesome ore till it was fairly "sweet," triturated it, and having reduced it with water to about the consistency of invalid's gruel, put it into a little berdan pan made from a "camp oven," which I had used for treating small quantities of concentrates, and from time to time drove a spray of mercury, wherein a small amount of zinc had been dissolved, into the pasty mass by means of a steam jet, added about half an ounce of sulphuric acid and kept the pan revolving for several hours. The result was an unusually successful amalgamation and consequent extraction—over ninety per cent.

Steam—or to use the scientific term, hydro-thermal action—has played such an important part in the deposition of metals that I cannot but think that under educated intelligence it will prove a powerful agent in their extraction. About fourteen years ago I obtained some rather remarkable results from simply boiling auriferous ferro-sulphides in water. There is in this alone an interesting, useful, and profitable field for investigation and experiment.

The most scientific and perfect mode of gold extraction (when the conditions are favourable) is lixiviation by means of chlorine, potassium cyanide, or other aurous solvent, for by this means as much as 98 per cent of the gold contained in suitable ores can be converted into its mineral salt, and being dissolved in water, re-deposited in metallic form for smelting; but lode stuff containing much lime would not be suitable for chlorination, or the presence of a considerable proportion of such a metal as copper, particularly in metallic form, would be fatal to success, while cyanide of potassium will also attack metals other than gold, and hence discount the effect of this solvent.

The earlier practical applications of chlorine to gold extraction were known as Mears' and Plattner's processes, and consisted in placing the material to be operated on in vats with water, and introducing chlorine gas at the bottom, the mixture being allowed to stand for a number of hours, the minimum about twelve, the maximum forty-eight. The chlorinated water was then drawn off containing the gold in solution which was deposited as a brown powder by the addition of sulphate of iron.

Great improvements on this slow and imperfect method have been made of late years, among the earlier of which was that of Messrs. Newbery and Vautin. They placed the pulp with water in a gaslight revolving cylinder, into which the chlorine was introduced, and atmospheric air to a pressure of 60 lb. to the square inch was pumped in. The cylinder with its contents was revolved for two hours, then the charge was withdrawn and drained nearly dry by suction, the resultant liquid being slowly filtered through broken charcoal on which the chloride crystals were deposited, in appearance much like the bromo-chlorides of silver ore seen on some of the black manganic oxides of the Barrier silver mines. The charcoal, with its adhering chlorides, was conveyed to the smelting-house and the gold smelted into bars of extremely pure metal. Messrs. Newbery and Vautin claimed for their process decreased time for the operation with increased efficiency.

At Mount Morgan, when I visited that celebrated mine, they were using what might be termed a composite adaptation process. Their chlorination works, the largest in the world, were putting through 1500 tons per week. The ore as it came from the mine was fed automatically into Krom roller mills, and after being crushed and sifted to regulation gauge was delivered into trucks and conveyed to the roasting furnaces, and thence to cooling floors, from which it was conveyed to the chlorinating shed. Here were long rows of revolving barrels, on the Newbery-Vautin principle, but with this marked difference, that the pressure in the barrel was obtained from an excess of the gas itself, generated from a charge of chloride of lime and sulphuric acid. On leaving the barrels the pulp ran into settling vats, somewhat on the Plattner plan, and the clear liquid having been drained off was passed through a charcoal filter, as adopted by Newbery and Vautin. The manager, Mr. Wesley Hall, stated that he estimated cost per ton was not more than 30s., and he expected shortly to reduce that when he began making his own sulphuric acid. As he was obtaining over 4 oz. to the ton the process was paying very well, but it will be seen that the price would be prohibitive for poor ores unless they could be concentrated before calcination.

The Pollok process is a newer, and stated to be a cheaper mode of lixiviation by chlorine. It is the invention of Mr. J. H. Pollok, of Glasgow University, and a strong Company was formed to work it. With him the gas is produced by the admixture of bisulphate of sodium (instead of sulphuric acid, which is a very costly chemical to transport) and chloride of lime. Water is then pumped into a strong receptacle containing the material for treatment and powerful hydraulic pressure is applied. The effect is stated to be the rapid change of the metal into its salt, which is dissolved in the water and afterwards treated with sulphate of iron, and so made to resume its metallic form.

It appears, however, to me that there is no essential difference in the pressure brought to bear for the quickening of the process. In each case it is an air cushion, induced in the one process by the pumping in of air to a cylinder partly filled with water, and in the other by pumping in water to a cylinder partly filled with air.

The process of extracting gold from lode stuff and tailings by means of cyanide of potassium is now largely used and may be thus briefly described:—It is chiefly applied to tailings, that is, crushed ore that has already passed over the amalgamating and blanket tables. The tailings are placed in vats, and subjected to the action of solutions of cyanide of potassium of varying strengths down to 0.2 per cent. These dissolve the gold, which is leached from the tailings, passed through boxes in which it is precipitated either by means of zinc shavings, electricity, or to the precipitant. The solution is made up to its former strength and passed again through fresh tailings. When the tailings contain a quantity of decomposed pyrites, partly oxidised, the acidity caused by the freed sulphuric acid requires to be neutralised by an alkali, caustic soda being usually employed.

When "cleaning up," the cyanide solution in the zinc precipitating boxes is replaced by clean water. After careful washing in the box, to cause all pure gold and zinc to fall to the bottom, the zinc shavings are taken out. The precipitates are then collected, and after calcination in a special furnace for the purpose of oxidising the zinc, are smelted in the usual manner.

The following description of an electrolytic method of gold deposition from a cyanide solution was given by Mr. A. L. Eltonhead before the Engineers' Club of Philadelphia.

A description of the process is as follows: — "The ore is crushed to a certain fineness, depending on the character of the gangue. It is then placed in leaching vats, with false bottoms for filtration, similar to other leaching plants. A solution of cyanide of potassium and other chemicals of known percentage is run over the pulp and left to stand a certain number of hours, depending on the amount of metal to be extracted. It is then drained off and another charge of the same solution is used, but of less strength, which is also drained. The pulp is now washed with clean water, which leaches all the gold and silver out, and leaves the tailings ready for discharge, either in cars or sluiced away by water, if it is plentiful.

"The chemical reaction of cyanide of potassium with gold is as follows, according to Elsner: —

$2Au + 4KCy + O + H_2O = 2KAuCy_2 + 2KHO$.

"That is, a double cyanide of gold and potassium is formed.

"All filtered solutions and washings from the leaching vats are saved and passed through a precipitating 'box' of novel construction, which may consist either of glass, iron or wood, and be made in any shape, either oval, round, or rectangular — if the latter, it will be about 10 ft. long, 4 ft. wide and 1 ft. high — and is partitioned off lengthwise into five compartments. Under each partition, on the inside or bottom of the 'box,' grooves may be cut a quarter-to a half-inch deep, extending parallel with the partitions to serve as a reservoir for the amalgam, and give a rolling motion to the solution as it passes along and through the four compartments. The centre compartment is used to hold the lead or other suitable anode and electrolyte.

"The anode is supported on a movable frame or bracket, so it may be moved either up or down as desired, it being worked by thumbscrews at each end.

"The electrolyte may consist of saturated solutions of soluble alkaline metals and earth. The sides or partitions of each compartment dip into the mercury, which must cover the 'box' evenly on the bottom to the depth of about a half-inch.

"Amalgamated copper strips or discs are placed in contact with the mercury and extended above it, to allow the gold and silver solution of cyanide to come in contact.

"The electrodes are connected with the dynamo; the anode of lead being positive and the cathode of mercury being negative. The dynamo is started, and a current of high amperage and low voltage is generated, generally 100 to 125 amperes, and with sufficient pressure to decompose the electrolyte between the anode and the cathode.

"As the gas is generated at the anode, a commotion is created in the liquid, which brings a fresh and saturated solution of electrolyte between the electrodes for electrolysis, and makes it continuous in its action.

"The solution of double cyanide of gold, silver, and potassium, which has been drained from the leaching vats, is passed over the mercury in the precipitating 'box' when the decomposition of the electrolyte by the electric current is being accomplished, the gold and silver are set free and unite with the mercury, and are also deposited on the plates or discs of copper, forming amalgam, which is collected and made marketable by the well known and tried methods. The above solution is regenerated with cyanide of potassium by the setting free of the metals in the passage over the 'box.'

"In using this solution again for a fresh charge of pulp, it is reinforced to the desired percentage, or strengthened with cyanide of potassium and other chemicals, and is always in good condition for continuing the operation of dissolving.

"The potassium acting on the water of the solution creates nascent hydrogen and potassium hydrate; the nascent hydrogen sets free the metals (gold and silver), which are precipitated into the mercury and form amalgam, leaving hydrocyanic acid; this latter combines with the potassium hydrate of the former reaction, thus forming cyanide of potassium. There are other reactions for which I have not at present the chemical formulas.

"As the solution passes over the mercury, the centre compartment of the 'box' is moved slowly longitudinally, which spreads the mercury, the solution is agitated and comes in perfect contact with the mercury, as well as the amalgamated plates or discs of copper, ensuring a perfect precipitation.

"It is not always necessary to precipitate all the gold and silver from the solution, for it is used over and over again indefinitely; but when it is required, it can be done perfectly and cheaply in a very short time.

"No solution leached from the pulp, containing cyanide of potassium, gold and silver, need be run to waste, which is in itself an enormous saving over the use of zinc shavings when handling large quantities of pulp and solution.

"Some of the advantages the electro-chemical process has over other cyanide processes are: Its cleanliness, quickness of action, cheapness, and large saving of cyanide of potassium by regeneration; not wasting the solutions, larger recovery of the gold and silver from the solutions; the cost of recovery less; the loss of gold, silver, and cyanide of potassium reduced to a minimum; the use of caustic alkali in such quantity as may be desired to keep the cyanide solution from being destroyed by the solidity of the pulp, and also sometimes to give warmth, as a warm cyanide solution will dissolve gold and silver quicker than a cold one. These caustic alkalies do not interfere with or prevent the perfect precipitation of the metals. The bullion recovered in this process is very fine, while the zinc-precipitated bullion is only about 700 fine.

"The gold and silver is dissolved, and then precipitated in one operation, which we know cannot be done in the 'chlorination process'; besides, the cost of plant and treatment is much less in the above-described process.

"The electro-chemical process, which I have hastily sketched will, I think, be the future cheap method of recovering fine or flour gold from our mines and waste tailings or ore dumps.

"Without going into details of cost of treatment, I will state that with a plant of a capacity of handling 10,000 tons of pulp per month, the cost should not exceed 8s. per ton, but that may be cheapened by labour-saving devices. There being no expensive machinery, a plant could be very cheaply erected wherever necessary."

CHAPTER VIII

CALCINATION OR ROASTING OF ORES

The object of calcining or roasting certain ores before treatment is to dissipate the sulphur or sulphides of arsenic, antimony, lead, etc., which are inimical to treatment, whether by ordinary mercuric amalgamation or lixivination. The effect of the roasting is first to sublimate and drive off as fumes the sulphur and a proportion of the objectionable metals. What is left is either iron oxide, "gossan," or the oxides of the other metals. Even lead can thus be oxidised, but requires more care as it melts nearly as readily as antimony and is much less volatile. The oxides in the thoroughly roasted ore will not amalgamate with mercury, and are not acted on by chlorine or cyanogen.

To effect the oxidation of sulphur, it is necessary not only to bring every particle of sulphur into contact with the oxygen of the air, but also to provide adequate heat to the particles sufficient to raise them to the temperature that will induce oxidation. No appreciable effect follows the mere contact of air with sulphur particles at atmospheric temperature; but if the particles be raised to a temperature of 500 degrees Fahr., the sulphur is oxidised to the gaseous sulphur dioxide. The same action effects the elimination of the arsenic and antimony associated with gold and silver ores, as when heated to a certain constant temperature these metals readily oxidise.

The science of calcination consists of the method by which the sulphide ores, having been crushed to a proper degree of fineness, are raised to a sufficient temperature and brought into intimate contact with atmospheric air.

It will be obvious then that the most effective method of roasting will be one that enables the particles to be thoroughly oxidised at the lowest cost in fuel and in the most rapid manner.

The roasting processes in practical use may be divided into three categories:

First or A Process.—Roasting on a horizontal and stationary hearth, the flame passing over a mass of ore resting on such hearth. In order to expose the upper surface of the ore to contact with air the material is turned over by manual labour. This furnace of the reverberatory type is provided with side openings by which the turning over of the ore can be manually effected, and the new ore can be charged and afterwards withdrawn.

Second or B Process.—Roasting in a revolving hearth placed at a slight incline angle from the horizontal. The furnace is of cylindrical form and is internally lined with refractory material. It has projections that cause the powdered ore to be lifted above the flame, and, at a certain height, to fall through the flame and so be rapidly raised to the temperature required to effect the oxidation of the oxidisable minerals which it is desired to extract.

The rate, or speed, of revolution of this revolving furnace obviously depends upon the character of the ore under treatment; it may vary from two revolutions per minute down to one revolution in thirty minutes. Any kind of fuel is available, but that of a gaseous character is stated to be by far the most efficient.

Any ordinary cylinder of a length of 25 ft., and a diameter of 4 ft. 6 in., inclined 1 ft. 6 in. in its length, will calcine from 24 to 48 tons per diem.

Another form of rotating furnace is one in which the axis is horizontal. It is much shorter than the inclined type, and the feeding and removal of the ore is effected by the opening of a retort lid door provided at the side of the furnace. Openings provided at each end of the furnace permit the passage of the flame through it, and the revolution of the furnace turns over the powdered ore and brings it into more or less sustained contact with the oxidising flame. The exposure of the ore to this action is continued sufficiently long to ensure the more or less complete oxidation of the ore particles.

Third or C Process.—In this process the powdered ore is allowed to fall in a shower from a considerable height, through the centre of a vertical shaft up which a flame ascends; the powdered ore in falling through the flame is heated to an oxidising temperature, and the sulphides are thus depleted of their sulphur and become oxides.

Another modification of this direct fall or shaft furnace is that in which the fall of the ore is checked by cross-bars or inclined plates placed across the shaft; this causes a longer oxidising exposure of the ore particles.

When the sulphur contents of pyritous ores are sufficiently high, and after the ore has been initially fired with auxiliary carbonaceous fuel, it is unnecessary, in a properly designed roasting furnace, to add fuel to the ore to enable the heat for oxidation to be obtained. The oxidation or burning of the sulphur will provide all the heat necessary to maintain the continuity of the process. The temperature necessary for effecting the elimination of both sulphur and arsenic is not higher than that equivalent to dull red heat; and provided that there is a sufficient mass of ore maintained in the furnace, the potential heat resulting from the oxidation of the sulphur will alone be adequate to provide all that is necessary to effect the calcination.

TYPES OF FURNACES OF THE DIFFERENT CLASSES THAT ARE IN ACTUAL USE.
"A" OR REVERBERATORY CLASS.

The construction of this furnace has already been sufficiently described. If the roasting is performed in a muffle chamber, the arrangement employed by Messrs. Leach and Neal, Limited, of Derby, and designed by Mr. B. H. Thwaite, C.E., can be advantageously employed in this furnace, which is fired with gaseous fuel. The sensible heat of the waste gases is utilised to heat the air employed for combustion; and by a controllable arrangement of combustion, a flame of over 100 feet in length is obtained, with the result that the furnace from end to end is maintained at a uniform temperature. By this system, and with gaseous fuel firing, a very considerable economy in fuel and in repairs to furnace, and a superior roasting effect, have been obtained.

Where the ordinary reverberatory hearth is fired with solid coal from an end grate, the temperature is at its maximum near the firing end, and tails off at the extreme gas outlet end. The ores in this furnace should therefore be fed in at the colder end of the hearth and be gradually worked or "rabbled" forward to the firing end.

One disadvantage of the reverberatory furnace is the fact that it is impossible to avoid the incursion of air during the manual rabbling action, and this tends to cool the furnace.

The cost of roasting, to obtain the more or less complete oxidation, or what is known in mining parlance as a "sweet roast" (because a perfectly roasted ore is nearly odourless) varies considerably, the variation depending of course upon the character of the ore and the cost of labour and fuel.

There are several modifications of the reverberatory furnace in use, designed mechanically to effect the rabbling. One of the most successful is that known as the Horse-shoe furnace. In plan the hearth of the furnace resembles a horse-shoe.

The stirring of the ore over the hearth is effected by means of carriages fixed in the centre of the furnace and having laterally projecting arms, carrying stirrers, that move along the hearth and turn over the pulverised ore.

In operation, half the carriages are traversing the furnace, and half are resting in the cooling space, so that a control over the temperature of the stirrers is established.

This furnace is stated to be more economical in labour than other mechanically stirred reverberatory furnaces, and there is also said to be an economy in fuel.

Usually the mechanical stirring furnaces give trouble and should be avoided, but the horse-shoe type possesses qualifications worthy of consideration.

"B." — THE REVOLVING CYLINDER FURNACE.

Of these the best known to me are: The Howell-White, the Bruckner, the Thwaite-Denny, and the Molesworth.

The Bruckner is a cylinder, turning on the horizontal axis and carried by four rollers.

The batch of ore usually charged into the two charging hoppers weighs about four tons. When the two charging doors are brought under the hopper mouth, the contents of the hopper fall directly into the cylinder.

The ends or throats of the furnace are reduced just sufficiently to allow the flame evolved from a grated furnace to pass completely through the cylinder.

A characteristic size for this Bruckner furnace is one having a length of 12 feet and a diameter of 6 feet. A furnace of this capacity will have an inclusive weight (iron and brickwork) of 15 tons.

The time of operation, with the Bruckner, will vary with the character of the ore under treatment and the nature of the fuel employed. Four hours is the minimum and twelve hours should be the maximum time of operation.

By the addition of common salt with the batch of ore, such of its constituents as are amenable to the action of chlorine are chlorinated as well as freed from sulphur.

Where the ore contains any considerable quantity of silver which should be saved, the addition of the salt is necessary as the silver is very liable to become so oxidised in the process of roasting as to render its after treatment almost impossible. I know a case in point where an average of nearly five ounces of silver to the ton, at that time worth 30s., was lost owing to ignorance on this subject. Had the ore been calcined with salt, NaCl, the bulk of this silver would have been amalgamated and thus saved. It was the extraordinary fineness of the gold saved by amalgamation as against my tests of the ore by fire assay that put me on the track of a most indefensible loss.

The Howell-White Furnace. — This furnace consists of a cast iron revolving cylinder, averaging 25 feet in length and 4 ft. 4 in. in diameter, which revolves on four friction rollers, resting on truck wheels, rotated by ordinary gearing.

The power required for effecting the revolution should not exceed four indicated horse-power.

The cylinder is internally lined with firebrick, projecting pieces causing the powdered ore to be raised over the flame through which it showers, and is thereby subjected to the influence of heat and to direct contact oxidation.

The inclination of the cylinder, which is variable, promotes the gradual descension of the ore from the higher to the lower end. It is fed into the upper end, by a special form of feed hopper, and is discharged into a pit at the lower end, from which the ore can be withdrawn at any time.

The gross weight of the furnace, which is, however, made in segments to be afterwards bolted together, is some ninety to one hundred tons.

The furnace is fired with coal on a grated hearth, built at the lower end; it is more economical both in fuel and in labour than an ordinary reverberatory furnace.

The Thwaite-Denny Revolving Furnace.—This new type of furnace, which is fired with gaseous fuel, is stated to combine the advantages of the Stetefeldt, the Howell-White, and the Bruckner.

It is constructed as follows:—Three short cylinders, conical in shape and of graduated dimensions, are superimposed one over the other, their ends terminating in two vertical shafts of brickwork, by which the three cylinders are connected. The powdered ore is fed into the uppermost cylinder and gravitates through the series. The highest cylinder is the largest in diameter, the lowest the smallest.

The gas flame, burnt in a Bunsen arrangement, enters the smallest end of the lowest cylinder and passes through it; then returns through the series and the ore is reduced by the expulsion of its sulphur, arsenic, etc., as it descends from the top to the bottom. The top cylinder is made larger than the one below it and the middle cylinder is made larger than the lowest one in proportion to the increased bulk of gases and ore.

The powdered ore in descending through the cylinders is lifted up and showers through the flame, falling in its descent a distance of over 1000 feet. By the time it reaches the bottom the ore is thoroughly roasted.

Provision is made for the introduction of separate supplies of air and gas into each cylinder; this enables the oxidising treatment to be controlled exactly as desired so as to effect the best results with all kinds of ore. Each cylinder is driven from its own independent gearing, and the speed of each cylinder can be varied at will.

The output of this type of furnace, the operations of which appear to be more controllable than those of similar appliances, depends, of course, upon the nature of the ore, but may be considered to range within the limits of twelve to fifty tons in twenty-four hours, and the cost of roasting will vary from 2s. 6d. to 4s. per ton, depending upon the quality of ore and of fuel.

The gaseous fuel generating system permits not only the absolute control over the temperature in the furnace, but the use of the commonest kinds of coal, and even charcoal is available.

The power required to drive the Thwaite-Denny furnace is four indicated horse-power.

The Molesworth Furnace also is a revolving cylindrical appliance, which, to say the least of it, is in many respects novel and ingenious. It consists of a slightly cone-shaped, cast-iron cylinder about fourteen feet long, the outlet end being the larger to allow for the expansion of the gases. Internal studs are so arranged as to keep the ore agitated; and spiral flanges convey it to the outlet end continually, shooting it across the cylinder. The cylinder is encased in a brick furnace. The firing is provided from *outside*, the inventor maintaining that the products of combustion are inimical to rapid oxidisation, to specially promote which he introduces an excess of oxygen produced in a small retort set in the roof of the furnace and fed from time to time with small quantities of nitrate of soda and sulphuric acid. Ores containing much sulphur virtually calcine themselves. I have seen this appliance doing good work. The difficulties appeared to be principally mechanical.

There are other furnaces which work with outside heat, but I have not seen them in action.

"C."—SHAFT TYPE OF FURNACE

In one form of this furnace, instead of allowing the ore to descend in a direct clear fall the descent is impeded by inclined planes placed at different levels in the height of the shaft, the ore descending from one plane to the other.

The Stetefeldt Shaft Furnace.—Although very expensive in first cost, has many advantages. No motive power is required and the structure of the furnace is of a durable character. Its disadvantages are:—Want of control, and the occasionally imperfect character of the roasting originating therefrom.

Three sizes of Stetefeldt's furnaces are constructed:

The largest will roast from 40 to 80 tons per diem.

The intermediate will roast from 20 to 40 tons per diem.

The smallest will roast from 10 to 20 tons per diem.

A good furnace should bring down the sulphur contents even of concentrates so as to be innocuous to mercuric amalgamation. The sulphur left in the ore should never be allowed to exceed two per cent.

A forty per cent pyritous or other sulphide ore should be roasted in a revolving furnace in thirty to forty minutes, and without any auxiliary fuel.

For ordinary purposes a 40-foot chimney is adequate for furnace work; such a chimney four feet square inside at the base, tapering to 2' 6" at the summit, will require 12,000 red bricks, and 1500 fire-bricks for an internal lining to a height of 12 feet from the base of the chimney shaft.

When second-hand Lancashire or Cornish boiler flues are available, they make admirable and inexpensive chimneys. The advantage of wrought-iron or steel chimneys lies in the convenience of removal and erection. They should be made in sections of 20 feet long, three steel wire guy-ropes attached to a ring, riveted to a ring two-thirds of the height of the chimney, and attached to holdfasts driven into the ground; tightening couplings should be provided for each wire.

Flue dust depositing chambers should be built in the line of the flues between the furnace and the chimney; they consist simply of carefully built brick chambers, with openings to enable workmen to enter and rapidly clear away the deposited matters. The chambers, three or four times the cross sectional area of the chimney flue, and ten to twenty feet long, can be built of brickwork, set in cement; the walls are provided with a cavity, filled with sand or Portland cement, so that there will be no danger of the incursion of air. In all furnace work the greatest possible precautions should be taken to prevent the least cracking of either joints or bricks. It is surprising how much the inadequate draft of a good chimney is due to cracks or orifices in the flues; and therefore a competent furnace-man should see to it that his flues are thoroughly sound, and free from openings through which the air can enter.[*]

> [*] For full details of the most recent improvements in the cyanide process and in other methods of extraction, the reader is referred to Dr. T. K. Rose's "Metallurgy of Gold," third edition.

CHAPTER IX

MOTOR POWER AND ITS TRANSMISSION

It is unnecessary to describe methods by which power for mining purposes has been obtained—that is, up to within the last five years—beyond a general statement, that when water power has been available in the immediate locality of the mine, this cheap natural source of power has been called upon to do duty. Steam has been the alternative agent of power production applied in many different ways, but labouring under as many disadvantages, chief of which are lack of water, scarcity of fuel and cost of transit of machinery. Sometimes condensing steam-engines have been employed. For the generation of steam the semi-portable and semi-tubular have been the type of boiler that has most usually been brought into service. Needless to say, when highly mineralised mine water only is available the adoption of this class of boiler is attended with anything but satisfactory results.

Recently, however, there is strong evidence that where steam is the power agent to be employed the water-tube type of boiler is likely to be employed, and to the exclusion of all other forms of apparatus for the generation of steam. The advantages of this type, particularly the tubulous form (or a small water tube), made as it is in sections, offers unrivalled facilities for transport service. The heaviest parts need not exceed 3 cwt. in weight, and require neither heavy nor yet expensive brickwork foundations.

WATERLESS POWER.

The difficulties in finding water to drive a steam plant are often of such a serious character as to involve the abandonment of many payable mines; therefore, a motive power that does not require the aqueous agent will be a welcome boon.

It will be a source of gratification to many a gold-claim holder to know that practical science has enabled motive power to be produced without the necessity of water, except a certain very small quantity, which once supplied will not require to be renewed, unless to compensate for the loss due to atmospheric evaporation.

Any carbonaceous fuel, such as, say, lignite, coal, or charcoal, can be employed. The latter can be easily produced by the method described in the Chapter on "Rules of Thumb," or by building a kiln by piling together a number of trunks of trees, or fairly large-sized branches, cut so that they can be built up in a compact form. The pile, after being covered with earth, is then lighted from the base, and if there are no inlets for the air except the limited proportion required for the smouldering fire at the base, the whole of the timber will be gradually carbonised to charcoal of good quality, which is available for the waterless power plant.

The waterless power plant consists of two divisions: First, a gas generating plant; secondly, an internal combustion or gas engine in which the gas is burnt, producing by thermo-dynamic action the motive power required. The system known as the Thwaite Power Gas System is not only practically independent of the use of water, but its efficiency in converting fuel heat into work is so high that no existing steam plant will be able to compete with it.

The weight of raw timber, afterwards to be converted into charcoal, that will be required to produce an effective horse-power for one hour equals 7 lb.

If coal is the fuel 1 1/3 lb. per E.H.P. for one hour's run.

If lignite is the fuel 2 1/2 lb. per E.H.P. for one hour's run.

The plant is simple to work, and as no steam boiler is required the danger of explosions is removed. No expensive chimney is necessary for the waterless power plant.

Where petroleum oil can be cheaply obtained, say for twopence per gallon, one of the Otto Cycle Oil Engines, for powers up to 20 indicated horse-power, can be advantageously employed.

These engines have the advantage of being a self-contained power, requiring neither chimney nor steam boiler, and may be said to be a waterless power. The objection is the necessity to rely upon oil as fuel, and the dangers attending the storage of oil. A good oil engine should not require to use more than a pint of refined petroleum per indicated horse-power working for one hour.

Fortunately for the mining industry electricity, that magic and mysterious agency, has come to its assistance, in permitting motive power to be transmitted over distances of even as much as 100 miles with comparatively little loss of the original power energy.

Given, that on a coal or lignite field, or at a waterfall, 100 horse-power is developed by the combustion of fuel or by the fall of water driving a turbine, this power can be electrically transmitted to a mine or GROUP OF MINES, say 100 miles away, with only a loss of some 30 horse-power. For twenty miles the loss on transmission should not exceed 15 horse-power so that 70 and 85 horse-power respectively are available at the mines. No other system offers such remarkable efficiencies of power transmission. The new Multiphase Alternating Electric Generating and Power Transmission System is indeed so perfect as to leave practically no margin for improvement.

The multiphase electric motor can be directly applied to the stamp battery and ore-breaker driving-shaft and to the shaft of the amalgamating pans.

> APPROXIMATE POWER REQUIRED TO DRIVE THE MACHINERY OF A MINE. Rock breaker 10 effective horse-power Amalgamating pan 5 effective horse-power Grinding pan 6 effective horse-power Single stamp of 750 lb. dropping 90 times per minute 1.25 effective horse-power Settlers 4 effective horse-power Ordinary hoisting lift 20 effective horse-power Allow 10 per cent in addition for overcoming friction.

Besides this electrical distribution power, which should not cost more than three farthings per effective horse-power per hour, the electrical energy can be employed for lighting the drives and the shafts of the mine. The modern electrical mine lamps leave little to be desired. Also it is anticipated that once the few existing difficulties have been surmounted electric drilling will supplant all other methods.

Electric power can be employed for pumping, for shot firing, for hauling, and for innumerable purposes in a mine.

Electricity lends itself most advantageously to so many and varied processes, even in accelerating the influence of cyanide solutions on gold, and in effecting the magnetic influence on metallic particles in separating processes; while applied to haulage purposes, either on aerial lines or on tram or railroads, it is an immediate and striking success.

It is anticipated that in the near future the mines on the Randt, South Africa, will be electrically driven from a coalfield generating station located on the coalfields some thirty miles from Johannesburg. Such a plant made up of small multiples of highly efficient machines will enable mine-owners to obtain a reliable power to any extent at immediate command and at a reasonable charge in proportion to the power used. This wholesale supply of power will be a godsend to a new field, enabling the opening up to be greatly expedited; and no climatic difficulties, such as dry seasons, or floods, need interfere with the regular running of the machinery. The same system of power-generation at a central station is to be applied to supply power to the mines of Western Australia.

CHAPTER X

COMPANY FORMATION AND OPERATIONS

All the world over, the operation of winning from the soil and rendering marketable the many valuable ores and mine products which abound is daily becoming more and more a scientific business which cannot be too carefully entered into or too skilfully conducted. The days of the dolly and windlass, of the puddler, cradle, and tin dish, are rapidly receding; and mining, either in lode or alluvial working, is being more generally recognised as one of the exact sciences. In the past, mining has been carried on in a very haphazard fashion, to which much of its non-success may be attributed.

But the dawn of better days has arrived, and with the advent of schools of mines and technical colleges there will in future be less excuse for ignorance in this most important industry.

This chapter will be devoted to Company formation and working, in which mistakes leading to very serious consequences daily occur.

It is not necessary to go deeply into the question why, in the mining industry more than any other, it should be deemed desirable as a general rule to carry on operations by means of public Companies, but, as a matter of fact, few names can be mentioned of men who mine extensively single handed. Yet, risky as it is, mining can hardly be said to be more subject to unpreventable vicissitudes than, say, pastoral pursuits, in which private individuals risk, and often lose or make, enormous sums of money.

However, it is with Mining Companies we are now dealing, and with the errors made in the formation and after conduct of these Associations.

The initial mistake most often made is that sufficient working capital is not called up or provided in the floating of the Company. Promoters trust to get sufficient from the ground forthwith to ensure further development; the consequence being that, as nearly 99 per cent of mining properties require a very considerable expenditure of capital before permanent profits can be relied on, the inexperienced shareholders who started with inflated hopes of enormous returns and immediate dividends become disheartened and forfeit their shares by refusing to pay calls, and thus many good properties are sacrificed. In England, the companies are often floated fully paid-up, but the same initial error of providing too little money for the equipment and effective working of the mine is usually fallen into.

Again, far too many Companies are floated on the report of some self-styled mining expert, often a man, who, like the schoolmaster of the last century, has qualified for the position by failing in every other business he has attempted. These men acquire a few geological and mining phrases, and by more or less skilfully interlarding these with statements of large lodes and big returns they supply reports seductive enough to float the most worthless properties and cause the waste of thousands of pounds. But the trouble does not end here.

When the Company is to be formed, some lawyer, competent or otherwise, is instructed to prepare articles of association, rules, etc.; which, three times out of four, is accomplished by a liberal employment of scissors and paste. Such rules may, or may not, be suited to the requirements of the organisation. Generally no one troubles much about the matter, though on these rules depends the future efficient working of the Company, and sometimes its very existence.

Then Directors have to be appointed, and these are seldom selected because of any special knowledge of mining they may possess, but as a rule simply because they are large shareholders or prominent men whose names look well in a prospectus. These gentlemen forthwith engage a Secretary, usually on the grounds that he is the person who has tendered lowest, to provide office accommodation and keep the accounts; and not from any particular knowledge he has of the true requirements of the position.

The way in which some Directors contrive to spend their shareholders' money is humorously commented on by a Westralian paper which describes a great machinery consignment lately landed in the neighbourhood of the Boulder Kalgoorlie.

"It would seem as if the purchaser had been let loose blindfold in a prehistoric material-founder's old iron yard, and having bought up the whole stock, had shipped it off. The feature of the entire antediluvian show is the liberal allowance of material devoted to destruction. Massive kibbles, such as were used in coal mines half a century ago, are arranged alongside a winding engine, built in the middle of the century, and evidently designed for hauling the kibbles from a depth of 1000 feet. Nothing less than horse-power will stir the trucks for underground use, and their design is distinctly of the antique type. The engine is built to correspond — of a kind that might have served to raise into position the pillars of Baalbec, and the mass of metal in it fairly raises a blush to the iron cheek of frailer modern constructions. The one grand use to which this monster could be put would be to employ it as a kedge for the Australian continent in the event of it dragging its present anchors and drifting down south, but as modern mining machinery the whole consignment is worth no more than its value as scrap-iron, which in its present position is a fraction or two less than nothing."

Next, a man to manage the mine has to be obtained, and some one is placed in charge, of whose capabilities the Directors have no direct knowledge. Being profoundly ignorant of practical mining they are incompetent to examine him as to his qualifications, or to check his mode of working, so as to ascertain whether he is acting rightly or not. All they have to rely on are some certificates often too carelessly given and too easily obtained. Finally, quite a large proportion of the allottees of shares have merely applied for them with the intention of selling out on the first opportunity at a premium, hence they have no special interest in the actual working of the mine.

Now let us look at the prospects of the Association thus formed. The legal Manager or Secretary, often a young and inexperienced man, knows little more than how to keep an ordinary set of books, and not always that. He is quite ignorant of the actual requirements of the mine, or what is a fair price to pay for labour, appliances, or material. He cannot check the expenditure of the Mining Manager, who may be a rogue or a fool or both, for we have had samples of all sorts to our sorrow. The Directors are in like case. Even where the information is honestly supplied, they cannot judge whether the work is being properly carried out or is costing a fair price, and the Mining Manager is left to his own devices, with no one to check him nor any with whom he can consult in specially difficult cases. Thus matters drift to the almost certain conclusion of voluntary or compulsory winding up; and so many a good property is ruined, and promising mines, which have never had a reasonable trial, are condemned as worthless. But let us ask, would any other business, even such as are less subject to unforeseen vicissitudes than mining, succeed under similar circumstances?

It is now very generally agreed that to the profitable development of mining new countries, at all events, must look mainly for prosperity, while other industries are growing. Therefore, we cannot too seriously consider how we may soonest make our mines successful.

What is the remedy for the unsatisfactory state of affairs we have experienced? The answer is a more practical system of working from the inception. Although it may evoke some difference of opinion I consider it both justifiable and desirable that the State should take some oversight of mining matters, at all events in the case of public Companies. It would be a salutary rule that the promoters of any mining undertaking should, before they are allowed to place it on the market, obtain and pay for the services of a competent Government Mining Inspector, who need not necessarily be a Government officer, but might, like licensed surveyors, be granted a certificate of competency either by a School of Mines or by some qualified Board of Examiners. The certificate of such Inspector that the property was as represented, should be given before the prospectus was issued. It is arguable whether even further oversight might not be properly be taken by the State and the report of a qualified officer be compulsory that the property was reasonably worth the value placed upon it in the prospectus.

Probably it will be contended that such restrictions would be an undue interference with private rights, and the old aphorism about a fool and his folly will be quoted. There are doubtless fools so infatuated that if they were brayed in a ten hundred-weight stamp-battery the "foolishness that had not departed from them" would give a highly payable percentage to the ton. Yet the State in other matters tries by numerous laws to protect such from their folly. A man may not sell a load of wood without the certificate from a licensed weighbridge or a loaf of bread without, if required, having to prove its weight; and we send those to gaol who practise on the credulity and cupidity of fools by means of the "confidence trick." Why not, therefore, where interests which may be said to be national are involved, endeavour to ensure fair dealing?

Then with regard to the men who are to manage the mines, seeing that a man may not become captain or mate of a river steamboat without some certificate on competency, nor drive her engines before he has passed an examination to prove his fitness, surely it is not too much to say that the mine manager or engineer, to whose care are often confided the lives of hundreds of men, and the expenditure of thousands of pounds, should be required to obtain a recognised diploma to prove his qualifications. The examinations might be made comparatively easy at first, but afterwards, when by the establishment of Schools and Mines the facilities have been afforded for men to thoroughly qualify, the standard should be raised; and after a date to be fixed no man should be permitted to assume the charge of a mine or become one of its officers without a proper certificate of competency from some recognised School of Mines or Technical College. The effect of such a regulation would in a few years produce most beneficial results.

In New Zealand, whose "progressive" legislature I do not generally commend, they have, in the matter of mine management, at all events, taken a step in the right direction. There a mine manager, before he obtains his certificate, must have served at least two years underground, and has to pass through a severe examination, lasting for days, in all subjects relating to mining and machinery connected with mining. In addition, he must prove his capacity by making an underground survey, and then plotting his work. The examination is a stiff one, as may be judged from the fact that between 1886 and 1891, only 27 candidates passed. Then the conditions were made easier, and from that date to 1895, 19 passed. Of the 46 students who gained first-class honours, 30 have left for South Africa or Australia, in both of which countries New Zealand certificated men are held in high estimation.

But returning to the formation of the Company, care should be taken in appointing Directors that at least one member of the Board is selected on account of his special technical knowledge of mining, and others for their special business capacity. The ornamental men with high sounding names should not be required in legitimate ventures. Also, it is most important that the business Manager or Secretary should be a specially qualified man, who by experience has learned what are the requirements of a mine doing a certain amount of work, so that a proper check may be kept on the expenses. The more Companies such a Secretary has the better, as one qualified man can supervise a large staff of clerks, who would themselves be qualifying for similar work, and gaining a useful and varied experience of mining business. An office of this description having charge of a large number of mines is, in its way, a technical school, and lads trained therein would be in demand as mine pursers, a very responsible and necessary officer in a big mine.

With respect to the men to whom the actual mining and treatment of ores and machinery is committed the greatest mistakes of the past have been that too much has been required from one man, a combination not to be found probably in one man in a thousand. Such Admirable Crichtons are rare in any profession or business, and that of mining is no exception. Men who profess too much are to be distrusted. Your best men are they who concentrate their energies and intellects in special directions. The Mining Manager should, if possible, be chosen from men holding certificates of competency from some technical mining school and, of course, should, in addition, have some practical experience, not necessarily as Head Manager. He should understand practical mine surveying and calculation of quantities, be able to dial and plot out his workings, and prepare an intelligible plan thereof for the use of the Directors, and should understand sufficient of physics, particularly pneumatics and hydraulics, to ensure thoroughly efficient pumping operations without loss of power from unnecessarily heavy appliances. Any other scientific knowledge applicable to his business which he may have acquired will tell in his favour, but he must, above all things, be a thoroughly practical man. Such men will in time be more readily procurable, as boys who have passed through the various Schools of Mines will be sent to learn their business practically at the mines just as we now, having given a lad a course of naval instruction, send him to sea to learn the practical part of his life's work.

But, of course, more is wanted on a mine than a man who can direct the sinking of shafts, driving of levels, and stoping of the lode. Much loss and disappointment have resulted in the past from unsuitable, ineffective, or badly designed and erected machinery, whether for working the mine or treating the ores. To obviate this defect a first-class mining engineer is required.

Then, also, day by day we are more surely learning that mining in all its branches is a science, and that the treatment of ores and extraction of the metals is daily becoming more and more the work of the laboratory rather than of the rule-of-thumb procedure of the past. Every mine, whether it be of gold, silver, tin, copper, or other metal, requires the supervision of a thoroughly qualified metallurgist and chemist, and one who is conversant with the newest processes for the extraction of the metals from their ores and matrices.

It has then been stated that to ensure effective working each mine requires, in addition to competent directors, a business manager, mining-manager, and assistants, engineer, chemist, and metallurgist, with assistant assayers, etc., all highly qualified men. But it will be asked, how are many struggling mines in sparsely populated countries to obtain the services of all these eminent scientists? The reply is by co-operation. One of the most ruinous mistakes of the past has been that each little mining venture has started on an independent course, with different management, separate machinery, etc. Can it then be wondered at that our gold-mining is not always successful?

Under a co-operative system all that each individual mine would require would be a qualified, practical miner capable of opening and securing the ground in a miner-like manner, and a good working engineer; and in gold-mining, where the gold is free in its matrix, a professional amalgamator, or lixiviator. For the rest, half a dozen or more mines may collectively retain the services of a mine manager of high attainments as general inspector and superintendent, and the same system could be adopted with respect to an advising metallurgist and an engineer. For gold, as indeed for other metals, a central extracting works, where the ores could be scientifically treated in quantity, might be erected at joint cost, or might easily be arranged for as a separate business.

A very fruitful cause of failure is the fatuous tendency of directors and mine managers to adopt new processes and inventions simply because they are new. As an inventor in a small way myself, and one who is always on the watch for improved methods, I do not wish to discourage intelligent progress; but the greatest care should be exercised by those having the control of the money of shareholders in mining properties before adopting any new machinery or process.

We have seen, and unfortunately shall see, many a promising mining company brought to grief by this popular error. The directors of mining companies might, to use an American saying, "paste this in their hats" as a useful and safe aphorism. "LET OTHERS DO THE EXPERIMENTING; WE ARE WILLING TO PAY ONLY FOR PROVED IMPROVEMENTS." I can cordially endorse every word of the following extracts from Messrs. McDermott and Duffield's admirable little work, "Losses in Gold Amalgamation."

"Some directors of mining companies are naturally inclined to listen to the specious promises of inventors of novel processes and new machinery, forgetting their own personal disadvantage in any argument on such matters, and assuming a confidence in the logic of their own conclusions, while they ignore the fruitful experience of thousands of practical men who are engaged in the mining business. The repeated failures of directors in sending out new machinery to their mines ought by this time to be a sufficient warning against increasing risks that are at once natural and unavoidable, and to deter them from plunging their shareholders into experiments which, in ninety-nine cases out of a hundred, result in nothing but excessive and needless expenses.

"It is certain that new machines and new processes are, and will be, given attention by mining men in proportion to their probable merits; but the proper place for experiments is in a mill already as successful as under known processes it can be made. In a new enterprise, even when the expense of an experiment is undertaken by the inventor, the loss to the mine-owner in case of failure must be very great, both in time and general running expenses. Directors should not believe that a willingness to risk cash in proving an invention is necessarily any proof of value of the same; it is only a measure of the faith of the inventor, which is hardly a safe standard to risk shareholders' money by.

"The variety of modifications in approved processes ought at least to suggest the desirability of exhausting the known, before drawing on the unknown and purely speculative. It should also be borne in mind that what might appear at first sight to be new processes, and even new machinery, are, in fact, often nothing but old contrivances and plausible theories long ago exploded among practical men.

"Many mining companies have been ruined, without any reference to their mines, through men deciding on the reasonableness of new process and machinery who have no knowledge of the business in hand. It is assumed often, that if an inventor or manufacturer of new machinery will agree to guarantee success, or take no pay if not successful, the company takes no risk. In actual fact a whole year is wasted in most cases, failure spoils the reputation of the company, running expenses have continued, and further working capital cannot be raised, because all concerned have lost confidence by the failure to obtain returns promised. All this in addition to the regular, unavoidable risks of mining itself, which may, at any moment during the year lost, call for increased expenses and increased faith in ultimate success. To the mining man who makes money by the business, the natural risks of mining is all he will take; it is sufficient; and when he invests more money in machinery he takes good care that he takes no chances of either failure or delay.

"The following are rules which no mining company or individual mine-owner can afford to neglect.

"(1) The risk should be confined to mining. No body of directors is justified in taking a shareholder's money and investing it in new processes or machinery when the subscription was simply for a mining venture. Directors are invariably incapable of deciding whether a so-called improvement in machinery or process is really so or not, and the reasonable course is to follow established precedents.

"(2) The risk of selecting an incompetent manager should be reduced to minimum by taking a man with a successful record in the particular work to be done. The manager selected should be prohibited, as much as the directors, from experimenting with new methods or machinery. A really experienced man will require no check in this direction, as he will not risk ruining his reputation.

"(3) The only time for a company to experiment is when the mine is paying well by the usual methods, and the treasury is in a condition to speculate a little in possible improvements without jeopardising regular returns."

Probably this is the best place to insert another word of warning to directors who are not mining specialists, and also to investors in gold mining shares. Assays of auriferous lode material are almost invariably worthless as a guide in the real value of the stone in quantity. The one way to decide this is by battery treatment in bulk, and then only after many tons have been put through. The reason is obvious. First, the prospector or company promoter, if he knows it, is not in the least likely to pick the worst piece of stone in the heap for assay; and, secondly, even should the sample be selected with the sole object of getting a fair result, no living man can judge the value of a gold lode by the result of treatment of an ounce of stone. So when you see it stated that Messrs. Oro and Gildenstein, the celebrated assayers, have found that a sample of rock from the Golden Mint Mine, Golconda, assays at the rate of 2,546 oz. 13 dwt. and 21 gr. to the ton, and that there are thousands of tons of similar stone in sight, the statement should be received with due caution. The assay is doubtless correct, but the deductions therefrom are most misleading.

A few words of advice also to directors of mine-purchasing companies and syndicates, of which there are now so many in existence, may probably be found of value. It is not good policy as a general rule to buy entirely undeveloped properties, unless such have been inspected by your own man, who is both competent and trustworthy, and who should have indeed an interest in the profits. Large areas, although so popular in England, do not compensate for large bodies of payable ore; the most remunerative mine is generally one of comparatively small area, but containing a large lode formation of payable but often low grade, ore.

It is worse still, of course, to buy a practically worked out mine, though this too is sometimes done. It must be remembered that mining, though often so profitable, is nevertheless a destructive industry, thus differing from agriculture, which is productive, and manufactures, which are constructive. Every ton of stone broken and treated from even the best gold mine in the world makes that mine the poorer by one ton of valuable material; thus, to buy a mining property on its past reputation for productiveness is, as a rule, questionable policy, unless you know there is sufficient good ore in sight to cover the purchase cost and leave a profit.

One of the greatest causes of non-success of gold-mining ventures, particularly when worked by public companies, is the lack of actual personal supervision, and hence, among other troubles, is that ultra-objectionable one—gold stealing from the mills, or, in alluvial mining, from the tail races. As to the former, the following appeared in 1893 in the London *Mining Journal*, and is, I think, worthy of the close consideration of mine directors in all parts of the world:—

"No one that has not experienced the evil of gold thieving from reduction mills can have any idea of the pernicious element it is, and the difficulty, once that it has got 'well hold,' of rooting it out. It permeates every class of society in the district connected with the industry, and managers, amalgamators, assayers, accountants, aye, even bank officials, are 'all on the job' to 'get a bit' while there is an opportunity. To exterminate the hateful monster requires on the part of the mine proprietors combined, stern and drastic measures undertaken under the personal supervision of one or more of their directors, and in many instances necessitating the removal of the whole of the official staff."

The writer narrates how about twenty years ago he was led to suspect that in an Australian mine running forty head of stamps, in which he held a controlling interest, the owners were being defrauded of about a fourth of the gold really contained in the ore, and the successful steps taken to check the robbery.

"We first of all dispensed with the services of the general manager, and then issued the following instructions to the mine and mill managers, I remaining at the mine to see them carried out until I substituted a practical local man as agent, who afterwards carried on the work most efficiently:—

"(a) Both of these officials to keep separate books and accounts; in other words, to be distinct departments.

"(b) The ore formerly was all thrown together and put through the mill. I subdivided it into four classes, A, B, C, and D, representing deep levels north and upper levels north, deep levels south and upper levels south, and allotted to each class ten heads of stamps at the mill.

"(c) The mine manager to try three prospects, forenoon and afternoon of each day, from the dumps of each of the four classes and record in a book to be kept for that purpose the estimated mill yield of each one.

"(d) The mill manager was required to do the same at the mill and keep his record.

"(e) There were four underground bosses in each shift, twelve in all. I had a book fixed at the top of the shaft in which I required each of these men, at the expiry of every shift, to record any change in the faces of the quartz and particularly in regard to quality.

"(f) Having divided the ore into four classes I instructed the amalgamators, of which there were two in each shift, six in all, that I required the amalgam from each to be kept separate, with the object of ascertaining what each part of the mine produced.

"(g) I procured padlocks for the covering boards of the mercury tables and gave the keys to the amalgamators with instructions that they were not to hand them over to any one except the exchange shift without my written authority, and instructed them that they should clean down the plates every three hours, and after cleaning down the amalgam, buckets to be placed in the cleaning room, which I instructed to be kept locked and the key in charge of the watchman night and day.

"(h) The whole of the amalgam taken from the plates during each twenty-four hours to be cleaned and squeezed by the two amalgamators on duty every forenoon at nine o'clock in the presence of the mill manager, who should weigh each lot and enter it in a book to be kept for the purpose, and the entry to be signed by the mill manager and both amalgamators as witnesses.

"(i) Every alternate Friday the mortars (boxes) to be cleaned out; the work to be commenced punctually at eight A.M. by the six amalgamators in the presence of the mill manager, assisted by the three amalgam cleaning room watchmen and the four battery feeders on duty, prohibiting any of them from leaving until the cleaning up was finished, and the amalgam cleaned, squeezed and weighed, and the amount entered by the mill manager in the record look and attested by the amalgamators.

"I think the intelligent readers (particularly those with a knowledge of the business) will see the drift of the above regulations, viz., for there to be any peculation the whole of the battery staff—fourteen in all—would have to participate in it, and the number was too many to keep a secret. Formerly the amalgam cleaning room was sacred to the mill manager, and on announcing to that official the new instructions he at once tendered his resignation in a tone of offended dignity, immediately followed by that of the mine manager. It is a significant fact that shortly afterwards these two officials purchased a large mill and other property at a cost of ten thousand pounds, and that the mine yielded for the following three years during which I was connected with it an average of over 17 dwt. to the ton, as against formerly 10 to 12 dwt.

"The reader must draw his own conclusions. I used to make it a practice to visit the mine daily and prospect the ore, and having the mine and mill managers' daily prospecting as a guide as well as my own, every man at the mill knew it was impossible for them to thieve without my detecting it; moreover, I made it a rule to discharge any of the mill employees that I discovered were interested in any small private claims.

"The crux of the whole thing is having a practical miner at the head of affairs, and it is impossible for him to thieve if the work is carried out in the manner I have described."

To bring the whole matter to a conclusion. It may be taken as a safe axiom that to make gold mining in the mine as distinct from mining on the Stock Exchange really profitable the same system of economy, of practical supervision, and scientific knowledge which is now adopted in all other businesses must be applied to the raising and extraction of the metal. Then, and not till then, will genuine mining take the place to which it is entitled amongst our industries.

CHAPTER XI

RULES OF THUMB

This chapter has been headed as above because a number of the rules and recipes given are simply practical expedients, not too closely scientific. My endeavour has been to supply practical and useful information in language as free from technicalities as possible, so as to adapt it to the ordinary miner, mill operator and prospector, many of whom have had no scientific training. Some of the expedients are original devices educed by what we are told is the mother of inventions; others are hints given by practical old prospectors who had met with difficulties which would be the despair of a man brought up within reach of forge, foundry, machine shop, or tradesmen generally. There are many highly ingenious and useful contrivances besides these I have given.

LIVING PLACES

The health of the prospector, especially in a new country, depends largely on his housing—in which particular many men are foolishly careless, for although they are aware that they will be camped out for long periods, yet all the shelter they rely on is a miserable calico tent, often without a "fly," while in some cases they sometimes even sleep on the wet, or dusty, ground. Such persons fully deserve the ill health which sooner or later overtakes them. A little forethought and very moderate ingenuity would render their camp comparatively healthy and comfortable.

In summer the tent is the hottest, and in winter the coldest of domiciles. The "pizie" or "adobie" hut, or, where practicable, the "dugout," are much to be preferred, especially the latter. "Pizie" or "adobie" is simply surface soil kneaded with water and either moulded between boards like concrete, to construct the walls, or made into large sun-dried bricks. Salt water should not be used, as it causes the wall to be affected by every change of weather. A properly constructed house of this material, where the walls are protected by overhanging eaves, are practically everlasting, and the former have been standing for centuries. There are buildings of pizie or adobie in Mexico, California and Australia which are as good as new, although the latter were built nearly a century ago.

Adobie dwellings are warm in winter and cool in summer, and can be kept clean and healthy by occasional coatings of lime whitewash.

The dugout is even more simple in construction. A cutting, say ten feet wide, is put into the base of a hill for say twelve feet until the back wall is, say, ten feet high, the sides starting from nothing to that height. The front and such portion as is required of the side walls are next constructed of pizie or rough stone, with mud mortar, and the roof either gabled or skillion of bough, grass, or reed thatch, and covered with pizie, over which is sometimes put another thin layer of thatch to prevent the pizie being washed away by heavy rain. Nothing can be more snug and comfortable than such a house, unless the cows, as Mark Twain narrates, make things "monotonous" by persistently tumbling down the chimney.

When the Burra copper mines were in full work in Australia, the banks of the Burra Creek were honeycombed like a rabbit warren with the "dugout homes" of the Cornish miners. The ruins of these old dugouts now extend for miles, and look something like an uncovered Pompeii.

When water is scarce and the tent has to be retained, much can be done to make the camp snug. I occupied a very comfortable camp once, of which my then partner, a Dane, was the architect. We called it "The Bungalow," and it was constructed as follows: First we set up our tent, 10 ft. by 8 ft., formed of calico, but lined with green baize, and covered with a well set fly.

Next we put in four substantial forked posts about 10 ft. high and 15 ft. apart, with securely fixed cross pieces, and on the top was laid a rough flat roof of brush thatch; the sides were then treated in the same way, but not so thickly, being merely intended as a breakwind.

The tent with its two comfortable bunks was placed a little to one side, the remaining space being used as a dining and sitting room all through the summer. Except in occasional seasons of heavy rain, when we were saved the trouble of washing our dishes, the tent was only used for sleeping purposes, and as a storehouse for clothes and perishable provisions. I have "dwelt in marble halls" since then, but never was food sweeter or sleep sounder than in the old bush bungalow.

A BUSH BED

To make a comfortable bush bedplace, take four forked posts about 3 ft. 6 in. long and 2 to 3 in. in diameter at the top; mark out your bedplace accurately and put a post at each corner, about 1 ft. in the ground. Take two poles about 7 ft. long, and having procured two strong five-bushel corn sacks, cut holes in the bottom corners, put the poles through, bringing the mouths of the sacks together, and secure them there with a strong stitch or two. Put your poles on the upright forked sticks, and you have a couch that even Sancho Panza would have envied. It is as well to fix stretchers or cross stays between the posts at head and foot.

In malarial countries, sleeping on the ground is distinctly dangerous, and as such districts are usually thickly timbered, the Northern Territory hammock is an admirable device, more particularly where mosquitoes abound.

NORTHERN TERRITORY HAMMOCK

This hammock, which is almost a standing bedplace when rigged, is constructed as follows:—To a piece of strong canvas 7 feet long and 2 1/2 feet wide, put a broad hem, say 3 1/2 inches wide at each end. Into this hem run a rough stick, about 2 feet 8 inches by 2 inches diameter. Round the centre of the stick pass a piece of strong three-quarter inch rope, 8 to 10 feet long and knot it, so as to leave a short end in which a metal eye is inserted. To each end of the two sticks a piece of quarter-inch lashing, about 6 feet long, is securely attached.

To make the mosquito covering take 18 feet of ordinary strong cheese cloth, and two pieces of strong calico of the same size as the canvas bed; put hems in the ends of the upper one large enough to take half-inch sticks, to all four extremities of which 8 feet of whipcord is to be attached. The calico forms the top and bottom of what we used to call the "meat safe," the sides being of cheese cloth. A small, flapped opening is left on the lower side. When once inside you are quite safe from mosquito bites.

To rig the above, two trees are chosen 7 to 8 feet apart, or two stayed poles can be erected if no trees are available. The bed is rigged about 3 feet from the ground by taking the rope round the trees or poles, and pulling the canvas taut by means of the metal eyelet. Then the lashings at the extremities of the sticks are fixed about 3 feet further up the trees and you have a bed something between a hammock and a standing bed. The mosquito net is fixed above the hammock in a similar manner, except that it does not require the centre stay.

An old friend of mine once had a rather startling experience which caused him to swear by the Northern Territory hammock. He was camped near the banks of a muddy creek on the Daly River, and had fortunately hung his "meat safe" about four feet high. The night was very dark, and some hours after retiring he heard a crash among his tin camp utensils, and the noise of some animal moving below him. Thinking his visitor was a stray "dingo," or wild dog, he gave a yell to frighten the brute away, and hearing it go, he calmly went to sleep again. Had he known who his caller really was, he would not have felt so comfortable. In the morning on the damp ground below, he found the tracks of a fourteen foot alligator, which was also out prospecting, but which, fortunately, had not thought of investigating the "meat safe."

PURIFYING WATER

There is not a more fertile disease distributor, particularly in a new country, than water. The uninitiated generally take it for granted that so long as water looks clear it is necessarily pure and wholesome; as a matter of fact the contrary is more usually the case, except in very well watered countries, and such, as a rule, are not those in which gold is most plentifully got by the average prospector. I have seen foolish fellows, who were parched with a long tramp, drink water in quantity in which living organisms could be seen with the naked eye, without taking even the ordinary precaution of straining it through a piece of linen. If they contracted hydatids, typhoid fever, or other ailments, which thin our mining camps of the strong, lusty, careless youths, who could wonder?

The best of all means of purifying water from organic substances is to boil it. If it be very bad, add carbon in the form of the charcoal from your camp fire. If it be thick, you may, with advantage, add a little of the ash also.

I once rode forty-five miles with nearly beaten horses to a native well, or rock hole, to find water, the next stage being over fifty miles further. The well was found, but the water in it was very bad; for in it was the body of a dead kangaroo which had apparently been there for weeks. The wretched horses, half frantic with thirst, did manage to drink a few mouthfuls, but we could not. I filled our largest billycan, holding about a gallon, slung it over the fire and added, as the wood burnt down, charcoal, till the top was covered to a depth of two inches. With the charcoal there was, of course, a little ash containing bi-carbonate of potassium. The effect was marvellous. So soon as the horrible soup came to the boil, the impurities coagulated, and after keeping it at boiling temperature for about half an hour, it was removed from the fire, the cinders skimmed out, and the water allowed to settle, which it did very quickly. It was then decanted off into an ordinary prospector's pan, and some used to make tea (the flavour of which can be better imagined than described); the remainder was allowed to stand all night, a few pieces of charcoal being added. In the morning it was bright, clear, and absolutely sweet. This experience is worth knowing as many a bad attack of typhoid and other fevers would be averted if practical precautions of this kind were only used.

TO OBTAIN WATER FROM ROOTS

The greatest necessity of animal life is water. There are, however, vast areas of the earth's surface where this most precious element is lamentably lacking, and such, unfortunately, is the case in many rich auriferous districts.

To the practical man there are many indications of water. These, of course, vary in different countries. Sometimes it is the herbage, but probably, the best of all is the presence of carnivorous animals and birds. These are never found far from water. In Australia the not over-loved wily old crow is a pretty sure indicator of water within reasonable distance — water may be extracted from the roots of the Mallee (*Eucalyptus dumosa* and *gracilis*) — the Box (*Eucalyptus hemiphloia*) and the Water Bush (*Hakea leucoptera*). To extract it the roots are dug up, cut into lengths of about a foot, and placed upright in a can; the lower ends being a few inches above the bottom. It is simply astonishing how much wholesome, if at times somewhat astringent, water may thus be obtained in a few hours, particularly at night.

Hakea leucoptera. "Pins and needles." — Maiden, in his work "Useful Native Plants of Australia," says: "In an experiment on a water-yielding *Hakea*, the first root, about half an inch in diameter and six or eight feet long, yielded quickly, and in large drops about a wine-glass full of really excellent water."

This valuable, though not particularly ornamental shrub (for it never attains to the dimensions of a tree), is found, to the best of my belief, in all parts of Australia, although it is said to be absent from West Australia. As to this I don't feel quite sure. I have seen it "from the centre of the sea" as far west as Streaky Bay, and believe I have seen it further West still. Considering the great similarity of much of the flora of South Africa to that of Australia, it is probable that some species of the water-bearing *Hakea* might be found there. It can readily be recognised by its acicular, needle-like leaves, and more particularly by its peculiarly shaped seed vessel, which resembles the pattern on an old-fashioned Indian shawl.

If the water found is too impure for drinking purposes and the trouble arises from visible animalculae only, straining through a pocket-handkerchief is better than nothing; the carbon filter is better still; but nothing is so effective as boiling. A carbon filter is a tube with a wad of compressed carbon inserted, through which the water is sucked, but as a rule clay-coloured water is comparatively innocuous, but beware of the bright, limpid water of long stagnant rock water-holes.

TO MAKE AN EFFECTIVE FILTER

Take a nail-can, keg, cask, or any other vessel, or even an ordinary wooden case (well tarred inside, if possible, to make it water-tight). Make a hole or several holes in the bottom, and set it over a tank or bucket. Into the bottom of the filter put (1) a few inches of washed broken stone; (2) about four inches of charcoal; (3) say three inches of clean coarse sand (if not to hand you can manufacture it by crushing quartz with your pestle and mortar), and (4) alternate layers of charcoal and sand until the vessel is half filled. Fill the top half with water, and renew from time to time, and you have a filter which is as effective as the best London made article. *But it is better to boil your water whether you filter afterwards or not.*

Clear the inside of the water-cask frequently, and occasionally add to the water a little Condy's fluid, as it destroys organic matter. A useful cement for stopping leaky places in casks is made as follows: Tallow 25 parts, lard 40 parts, sifted wood ash 25 parts. Mix together by heating, and apply with a knife blade which has just been heated.

CANVAS WATER BAGS

Are easily made, and are very handy for carrying small supplies of drinking-water when prospecting in a dry country; they have the advantage of keeping the water cool in the hottest weather, by reason on the evaporation. The mouthpiece is made of the neck of a bottle securely sewn in.

MEDICINE CASE

Medicine is also a matter well worthy of thought. The author's worst enemy would not call him a mollycoddle, yet he has never travelled in far wilds without carrying something in the way of medicine. First, then, on this subject, it cannot be too often reiterated that if common Epsom salts were a guinea an ounce instead of a penny the medicine would be valued accordingly, but it is somewhat bulky. What I especially recommend, however, is a small pocket-case of the more commonly known homeopathic remedies, "Mother tinctures," which are small, light, and portable, with a small simple book of instructions. Though generally an allopath in practice, I once saved my own life, and have certainly helped others by a little knowledge in diagnosing complaints and having simple homeopathic remedies at hand to be used in the first stages of what might otherwise have been serious illnesses.

PRODUCING FIRE

Every one has heard, and most believe, that fire may be easily produced by rubbing together two pieces of wood. I have seen it done by natives, but they seldom make use of the operation, which is generally laborious, preferring to carry lighted fire sticks for miles. I have never succeeded in the experiment.

Sometimes, however, it is almost a matter of life or death to be able to produce fire. The back of a pocket knife, or an old file with a fragment of flint, quartz, or pyrites struck smartly together over the remains of a burnt piece of calico, will in deft hands produce a spark which can be fanned to a glow, and so ignite other material, till a fire is produced.

Also it may not be generally known that he who carries a watch carries a "burning glass" with which he can, in clear weather, produce fire at will. All that is required is to remove the glass of your watch and carefully three parts fill it with water (salt or fresh). This forms a lens which, held steadily, will easily ignite any light, dry, inflammable substance.

When firearms are carried, cut a cartridge so that only about a quarter of the charge of powder remains. Damp some powder and rub it on a small piece of dry cotton cloth or well-rubbed brown paper. Push a loose pellet of this into the barrel, insert your half cartridge, fire at the ground, when the wad will readily ignite, and can be blown into flame.

TO COPY CORRESPONDENCE

The prospector is not usually a business man; hence in dealing with business men who, like Hamlet, are "indifferent honest," he frequently comes to grief through not having a copy of his correspondence. It is most desirable, therefore, either to carry a carbon paper duplicating book and a stylus, or by adding a little sugar to good ordinary black ink you may make a copying ink; then with the aid of a "yellow back" octavo novel, two pieces of board, and some ordinary tissue paper, you may take a copy of any letter you send.

TO PROVIDE A SIMPLE TELEGRAPHIC CODE

Buy a couple of cheap small dictionaries of the same edition, send one to your correspondent with an intimation that he is to read up or down so many words from the one indicated when receiving a message. Thus, if I want to say "Claim is looking well," I take a shilling dictionary, send a copy to my correspondent with the intimation that the real word is seven down, and telegraph—"Civilian looking weird;" this, if looked up in Worcester's little pocket dictionary, for instance, will read "Claim looking well." Any dictionary will do, so long as both parties have a copy and understand which is the right word. By arrangement this plan can be varied from time to time if you have any idea that your code can be read by others.

A SERVICEABLE SOAP

Wood ashes from the camp fire are boiled from day to day in a small quantity of water, and allowed to settle, the clear liquid being decanted off. When the required quantity of weak lye has been accumulated, evaporate by boiling, till a sufficient degree of strength has been obtained. Now melt down some mutton fat, and, while hot, add to the boiling lye. Continue boiling and stirring till the mixture is about the consistency of thick porridge, pour into any convenient flat vessel, and let it stand till cool. If you have any resin in store, a little powdered and added gradually to the melting tallow, before mixing with the lye, will stiffen your soap.

TO CROSS A FLOODED STREAM

Take a half-gallon, or larger, tin "billy can," enclose it in a strong cotton handkerchief or cotton cloth, knotting same over the lid, invert, and, taking the knot in the hand, you have a floating appliance which will sustain you in any water, whether you are a swimmer or not. The high silk hat of civilisation would act as well as the can, but these are not usually found far afield.

TO MAKE A HIDE BUCKET

At times when prospecting in an "incline" or "underlay" shaft, particularly where the walls of the lode are irregular, a hide bucket will be found preferable to an iron one. The mode of manufacture is as follows: Procure an ox hide, "green," if possible; if dry, it should be soaked until quite soft. Cut some thin strips of hide for sewing or lacing. Now shape a bag or pocket of size sufficient to hold about a hundredweight of stone, and by puncturing the edges with a knife, marline-spike, or other pointed tool, sew together; make a handle of twisted or pleated hide, and having filled your bucket with dry sand or earth let it stand till the whole is quite dry, when it will be properly distended and will maintain its shape until worn out.

TO MAKE A "SLUSH LAMP"

Where candles are scarce and kerosine is not, a "slush lamp" is a useful substitute. Take an old but sound quart tin pannikin, half fill it with sand or earth, and prepare a thin stick of pine, round which wrap a strip of soft cotton cloth. The stick should be about half an inch longer than the depth of the pannikin. Melt some waste fat, fill the pannikin therewith, push the stick down into the earth at the bottom, and you have a light, which, if not equal to the electric or incandescent gas burner, is quite serviceable. In Australia the soft velvety core of the "bottle brush," *Banksia marginata*, is often used instead of the cotton wick.

CHAPTER XII

RULES OF THUMB
MINING APPLIANCES AND METHODS
A TEMPORARY FORGE

What prospector has not at times been troubled for the want of a forge? To steel or harden a pick or sharpen a drill is comparatively easy, but there is often a difficulty in getting a forge. Big single action bellows are sometimes bought at great expense, and some ingenious fellows have made an imitation of the blacksmith's bellows by means of sheepskins and rough boards.

With inadequate material and appliances to hand, the following will be found easier to construct and more lasting when constructed. Only a single piece of iron is required, and, at a pinch, one could even dispense with that by using a slab of talcose material, roughly shaping a hearth therein and making a hole for the blast. First, construct a framing about the height of an ordinary smith's forge. This can made with saplings and bark, or better still, if available, out of an empty packing case about three feet square. Fill the frame or case with slightly damped earth and ram it tight, leaving the usual hollow hearth. Then form a chamber below the perforated hearth opening to the rear. Now construct a centrifugal fan, such as is used for the ventilation of shallow shafts and workings. Set this up behind the hearth and revolve by means of a wooden multiplying wheel. A piece of ordinary washing line rope, or sash line rope, well resined if resin can be got—but pitch, tar, or wax will do by adding a little fine dust to prevent sticking—is used as a belt. With very rough materials a handy man can thus make a forge that will answer ordinary requirements.—N.B. Do not use clay for your hearth bed unless you can get a highly aluminous clay, and can give it full time to dry before the forge fire is lit. Ordinary surface soil, not too sandy, acts well, if damped and rammed thoroughly. Of course, if you can get an iron nozzle for your blower the whole operation is simplified.

SIMPLE WAY OF MAKING CHARCOAL

Dig a pit 5 feet square by 3 feet deep and fill with fuel. After lighting, see that the pit is kept full. The hot embers will gradually sink to the bottom. The fuel should be kept burning fiercely until the pit seems almost full, when more fuel should be added, raising the heap about a foot above the level of the ground. The earth dug out of the pit should then be shovelled back over the burning mass. After leaving it to cool for 24 hours the pit will be found nearly full of charcoal. About one-quarter the weight of the dry fuel used should be recovered in charcoal.

ROUGH SMELTING ON THE MINE

Rough smelting on the mine is effected with a flux of borax, carbonate of soda, or, as I have often done, with some powdered white glass. When the gold is smelted and the flux has settled down quietly in a liquid state, the bulk of the latter may be removed, to facilitate pouring into the mould, by dipping an iron rod alternately into the flux and then into a little water, and knocking off the ball of congealed flux which adheres after each dip. This flux should, however, be crushed with a pestle and mortar and panned off, as, in certain cases, it may contain tiny globules of gold.

MISFIRES IN BLASTING

One of the most common sources of accident in mining operations is due either to carelessness or to the use of defective material in blasting. A shot misses, generally for one of two reasons; either the explosive, the cap, or the fuse (most often the latter), is inferior or defective; or the charging is incompletely performed. Sometimes the fuse is not placed properly in the detonator, or the detonator is not properly enclosed in the cartridge, or the fuse is injured by improper tamping. If several shots have been fired together, particularly at the change of a "shift," the men who have to remove the broken material may in so doing explode the missed charge. Or, more inexcusable still, men will often be so foolish as to try to clear out the drill hole and remove the missed cartridge. When a charge is known to have missed all that is necessary to do in order to discharge it safely is to remove a few inches of "tamping" from the top of the drill hole, place in the bore a plug of dynamite with cap and fuse attached, put an inch or two of tamping over it and fire, when the missed charge will also be exploded. Of course, judgment must be used and the depth of the drill taken into consideration. As a rule, miners use far more tamping than is at all requisite. The action of the charge will generally be found quite as effective with a few inches of covering matter as with a foot or more, while the exploding of misfire cartridges is rendered simple, as no removal of tamping is required before placing the top "plug" in case of misfire.

TO PREVENT LOSS OF RICH SPECIMENS IN BLASTING

When blasting the cap of a lode, particularly on rich shutes of gold, the rock is apt to fly, and rich specimens may be thrown far afield and so be lost. A simple way of avoiding this is to procure a quantity of boughs, which tie into loose bundles, placing the leafy parts alternately end for end. Before firing, pile these bundles over the blast and, if care is used, very few stones will fly. The same device may be used in wide shallow shafts.

A SIMPLE MODE OF RETORTING SMALL QUANTITIES OF AMALGAM

Clean your amalgam and squeeze it as hard as possible through strong calico or chamois leather. Take a large sound potato, cut off about a quarter from one end and scoop out a hole in the centre about twice as big as the ball of amalgam. Procure a piece of flat iron—an old spade will do as well as anything—insert the amalgam, and, having placed the potato, cut side downwards, thereon, put the plate of iron on the forge, heat up first gently, then stronger, till separation has taken place, when the gold will be found in a bright clean button on the plate and the mercury in fine globules in the potato, from which it can be re-collected by breaking up the partly or wholly cooked tuber under water in an enamelled or ordinary crockery basin.

TO RETORT SMALL QUANTITIES OF MERCURY FOR AMALGAMATING ASSAY TESTS

Get two new tobacco pipes similar in shape, with the biggest bowls and longest stems procurable. Break off the stem of one close to the bowl and fill the hole with well worked clay (some battery slimes make the best luting clay). Set the stemless pipe on end in a clay bed, and fill with amalgam, pass a bit of thin iron or copper wire beneath it, and bend the ends of the wire upwards. Now fit the whole pipe, bowl inverted, on to the under one, luting the edges of both well with clay. Twist the wire over the top with a pair of nippers till the two bowls are fitted closely together, and you have a retort that will stand any heat necessary to thoroughly distil mercury.

A SIMPLE MODE OF ASCERTAINING THE NOMINAL HORSE-POWER OF AN ENGINE

Multiply the internal diameter of the cylinder by itself and strike off the last figure of the quotient. The diameter is

    20" X 20" 20 ____ 400. The H.P. is 40.

The following rules will be found more professionally accurate from an engineering standpoint, though the term "horse-power" is not now generally employed.

To find the Nominal Horse-power. — For *non-condensing* engines: Multiply the square of the diameter of the cylinder in inches by 7 and divide the product by 80. For *condensing* engines: Multiply the square of the diameter of the cylinder in inches by 7 and divide the product by 200.

To find the Actual Horse-power of an engine, multiply the area of the cylinder in square inches by the average effective pressure in pounds per square inch, less 3 lb. per square inch as the frictional allowance, and also by the speed of the piston in feet per minute, dividing the product by 33,000, and the quotient will be the actual horse-power.

"SCALING" COPPER PLATES

To "scale" copper plates they may be put over a charcoal or coke fire to slowly sublimate the quicksilver. Where possible, the fireplace of a spare boiler can be utilised, using a thin red fire. After the entire evaporation of the quicksilver the plates should be slowly cooled, rubbed with hydrochloric acid, and put in a damp place overnight, then rubbed with a solution of sal ammoniac and nitre in equal parts, and again heated slowly over a red fire. They must not be allowed to get red hot; the proper degree of heat is indicated by the gold scale rising in blisters, when the plates should be taken from the fire and the gold scraped off. Any part of the plate on which the gold has not blistered should be again rubbed with the solution and fired. The gold scale should be collected in a glass or earthen dish and covered with nitric acid, till all the copper is dissolved, when the gold can be smelted in the usual way; but after it is melted corrosive sublimate should be put in the crucible till a blue flame ceases to be given off.

A Second Method

The simplest plan I know is to have a hole dug nine inches deep by about the size of the plate to be scaled; place a brick at each corner, and on each side, halfway between, get up a good fire; let it burn down to strong embers, or use charcoal, then place the plate on three bars of iron extending between the three pairs of bricks, have a strong solution of borax ready in which soak strips of old "table blanket," laying these over the plate and sprinkling them with the borax solution when the plate gets too hot. After a time the deposit of mercury and gold on the plate will assume a white, efflorescent appearance, and may then be readily parted from the copper.

Another Method

Heat the plate over an open fire, to drive off the mercury; after which, let it cool, and saturate with dilute sulphuric acid for three hours, or longer; then sprinkle over the surface a mixture of equal parts of common salt and sal ammoniac, and heat to redness; then cool, and the gold scale comes off freely; the scale is then boiled in nitric or sulphuric acid, to remove the copper, previous to melting. Plates may be scaled about once in six months, and will under ordinary circumstances produce about one ounce of clean gold for each superficial foot of copper surface employed. I always paint the back of the plate with a mixture of boiled oil and turpentine, or beeswax dissolved in turpentine, to prevent the acid attacking the copper.

HOW TO SUPPLY MERCURY TO MORTAR BOXES

I am indebted for the following to Mr. J. M. Drake, who, speaking of his experience on the Wentworth Mine, N.S.W., says:

"Fully 90 per cent of the gold is saved on the outside plates, only a small quantity remaining in the mortar. The plates have a slope of 2 in. to 1 ft. No wells are used, the amalgam traps saving any quicksilver which may leach off the plates. The quicksilver is added every hour in the mortar. The quantity is regulated by the mill manager in the following manner: Three pieces of wood, 8 in. wide by 12 in. long by 2 in. thick, have 32 holes 1 in. deep bored in each of them. These holes will just take a small 2 oz. phial. The mill manager puts the required quantity of quicksilver in each bottle and the batteryman empties one bottle in each mortar every hour; and puts it back in the hole upside down. Each block of wood lasts eight hours, the duration of one man's shift." This of course is for a 20-head mill with four mortars or "boxes."

I commend this as an excellent mode of supplying the mercury to the boxes or mortars. The quantity to be added depends on circumstances. A careless battery attendant will often put in too much or too little when working without the automatic feeder. I have known an attendant on suddenly awaking to the fact half through his shift, that he had forgotten to put in any mercury, to then empty into the stamper box two or three pounds weight; with what effect may be easily surmised.

HOW WATER SHOULD ENTER STAMPER BOXES

The following extract which relates to Californian Gold Mill practices is from Bulletin No. 6 of the California State Mining Bureau. I quite agree with the practice.

"The battery water should enter both sides of the mortar in an even quantity, and should be sufficient to keep a fairly thick pulp which will discharge freely through the grating or screen. About 120 cubic feet of water per ton of crushed ore may be considered an average, or 8 to 10 cubic feet per stamp per hour.

"Screens of different materials and with different orifices are used; the materials comprise wire cloth of brass or steel, tough Russian sheet iron, English tinned plate, and, quite recently, aluminium bronze. The 'aluminium bronze' plates are much longer lived than either of the other kinds, and have the further advantage that, when worn out, they can be sold for the value of the metal for remelting; these plates are bought and sold by the pound, and are said to contain 95 per cent of copper and 5 per cent of aluminium. Steel screens are not so much used, on account of their liability to rust."

I have had no experience with the aluminium bronze screen. I presume, however, that it is used only for mills where mercury is not put in the mortars, otherwise, it would surely become amalgamated. The same remark applies to brass wire cloth and tinned plate. Unless the metal of which they are composed will not readily amalgamate with mercury, I should be chary of using new screen devices. Mercury is a most insidious metal and is often found most unexpectedly in places in the battery where it should not be. Probably aluminium steel would be better than any substance mentioned. It would be hard, light, strong, and not readily corrodible. I am not aware if it has been tried.

Under the heading of "Power for Mills" the following is taken from the same source.

POWER FOR MILLS

"As the Pelton wheel seems to find the most frequent application in California, it may be convenient for millmen to have the following rule, applicable to these wheels:

"When the head of water is known in feet, multiply it by 0.0024147, and the product is the horse-power obtainable from one miner's inch of water.

"The power necessary for different mill parts is:

> For each 850lb. stamp, dropping 6 in. 95 times per minute, 1.33 h.-p. For each 750lb. stamp, dropping 6 in. 95 times per minute, 1.18 h.-p. For each 650lb. stamp, dropping 6 in. 95 times per minute, 1.00 h.-p. For an 8-inch by 10-inch Blake pattern rock-breaker 9.00 h.-p. For a Frue or Triumph vanner, with 220 revolutions per min. 0.50 h.-p. For a 4-feet clean-up pan, making 30 revolutions per min. 1.50 h.-p. For an amalgamating barrel, making 30 revolutions per min. 2.50 h.-p. For a mechanical batea, making 30 revolutions per min. 1.00 h.-p."

The writer has had small practical experience of the working of that excellent hydraulic motor, the Pelton wheel, but if by horse-power in the table given is meant nominal horse-power, it appears to be high. Working with 800 cwt. stamps, 80 blows a minute, one horse-power nominal will be found sufficient with any good modern engine, which has no further burden than raising the stamps and pumping the feed water. It is always well, however, particularly when providing engine power, to err on the right side, and make provision for more than is absolutely needed for actual battery requirements. This rule applies with equal potency to pumping engines.

TO AVOID LOSS IN CLEANING UP

The following is a hint to quartz mill managers with respect to that common source of loss of gold involved in the almost inevitable loss of mercury in cleaning up operations. I have known hundreds of pounds' worth of gold to be recovered from an old quartz mill site by the simple process of washing up the ground under the floor.

If you cannot afford to floor the whole of the battery with smooth concrete, at all events smoothly concrete the floor of the cleaning-up room, and let the floor slope towards the centre: where a sink is provided. Any lost mercury must thus find its way to the centre, where it will collect and can be panned off from time to time. Of course an underground drain and mercury trap must be provided.

IRON EXTRACTOR

When using self-feeders, fragments of steel tools are especially liable to get into the battery boxes or other crushing appliance where they sometimes cause great mischief. I believe the following plan would be a practicable remedy for this evil.

By a belt from the cam or counter shaft, cause a powerful electric magnet to extract all magnetic particles; then, by a simple ratchet movement, at intervals withdraw the magnet and drop the adhering fragments into a receptacle by automatically switching off the electric current. A powerful ordinary horseshoe magnet might probably do just as well, but would require to be re-magnetised from time to time.

TO SILVER COPPER PLATES

To silver copper plates, that is, to amalgamate them on the face with mercury, is really a most simple operation, though many batterymen make a great mystery of it. Indeed, when I first went into a quartz mill the process deemed necessary was not only a very tedious one, but very dirty also.

To amalgamate with silver, in fact, to silver-plate your copper without resort to the electro-plating bath, take any old silver (failing that, silver coin will do, but is more expensive), and dissolve it in somewhat dilute nitric acid, using only just sufficient acid as will assist the process. After some hours place the ball of amalgam in a piece of strong new calico and squeeze out any surplus mercury.

About an ounce of silver to the foot of copper is sufficient. To apply it on new plates use nitric acid applied with a swab to free the surface of the copper from oxides or impurities, then rub the ball of amalgam over the surface using some little force. It is always well when coating copper plates with silver or zinc by means of mercury to let them stand dry for a day or two before using, as the mercury oxidises and the coating metal more closely adheres.

Only the very best copper plate procurable should be used for battery tables; bad copper will always give trouble, both in the first "curing," and after treatment. It should not be heavily rolled copper, as the more porous the metal the more easily will the mercury penetrate and amalgamate. I cannot agree that any good is attained by scouring the plates with sand and alkalies, as recommended in some books on the subject; on the contrary, I prefer the opposite mode of treatment, and either face the plates with nitrate of silver and nitrate of mercury, or else with sulphate of zinc and mercury, in the form of what is called zinc amalgam. If mine water, which often contains a little free sulphuric acid, is being used, the latter plan is preferable.

The copper should be placed smoothly on the wooden table and secured firmly thereto by copper tacks. If the plate should be bent or buckled, it may be flattened by beating it with a heavy hammer, taking care to interpose a piece of inch-thick soft wood between hammer and plate.

To coat with mercury only, procure some nitrate of mercury. This is easily made by placing mercury in an earthenware bowl, pouring somewhat dilute nitric acid on it, and letting it stand till the metallic mercury is changed to a white crystal. Dense reddish-brown fumes will arise, which are injurious if breathed, so the operation should be conducted either in the open air, or where there is a draught.

Having your silvering solution ready, which is to be somewhat diluted with water, next take two swabs, with handles about 12 inches long, dip the first into a basin containing dilute nitric acid, and rub it rapidly over about a foot of the surface of the plate; the oxide of copper will be absolutely removed, and the surface of the copper rendered pure and bright; then take the other swab, wet with the dilute nitric of mercury, and pass it over the clean surface, rubbing it well in. Continue this till the whole plate has a coating of mercury. It may be well to go over it more than once. Now turn on the water and wash the plate clean, sprinkle with metallic mercury, rubbing it upwards until the plate will hold no more.

A basin with nitrate of mercury may be kept handy, and the plates touched up from time to time for a few days until they get amalgamated with gold, after which, unless you have much base metal to contend with, they will give no further trouble.

It must be remembered, however, that an excessive use of nitric acid will result in waste of mercury, which will be carried off in a milky stream with the water; and also that it will cause the amalgam to become very hard, and less active in attracting other particles of gold.

If you are treating the plate with nitrate of silver prepared as already mentioned, clean the plate with dilute nitric acid, rub the surface with the ball of amalgam, following with the swab and fairly rubbing in. It will be well to prepare the plate some days before requiring to use it, as a better adhesion of the silver and copper takes place than if mercury is applied at once.

To amalgamate with zinc amalgam, clean the copper plate by means of a swab, with fairly strong sulphuric acid diluted with water; then while wet apply the zinc-mercury mixture and well rub in. To prepare the zinc-amalgam, clip some zinc (the lining of packing cases will do) into small pieces and immerse them in mercury after washing them with a little weak sulphuric acid and water to remove any coating of oxide. When the mercury will absorb no more zinc, squeeze through chamois leather or calico (as for silver amalgam), and well rub in. The plate thus prepared should stand for a few days, dry, before using. If, before amalgamation with gold takes place, oxide of copper or other scum should rise on this plate a little very dilute sulphuric acid will instantly remove it.

Sodium and cyanide of potassium are frequently used in dressing-plates, but the former should be very sparingly employed, as it will often do more harm than good by taking up all sorts of base metals with the amalgam, and so presenting a surface which the gold will pass over without adhering to. Where water is scarce, and is consequently used over and over again, lime may be added to the pulp, or, if lime is not procurable, wood ashes may be used. The effect is two-fold; the lime not only tends to "sweeten" sulphide ores and keep the tables clean, but also causes the water to cleanse itself more quickly of the slimes, which will be more rapidly precipitated. When zinc amalgam is used, alkalies would, of course, be detrimental.

When no other water than that from the mine is available, difficulties often arise owing to the impurities it contains. These are various, but among the most common are the soluble sulphates, and sometimes free sulphuric acid evolved by the oxidisation of metallic sulphides. In the presence of this difficulty, do one of two things; either *utilise* or *neutralise*. In certain cases, I recommend the former. Sometime since I was treating, for gold extraction, material from a mine which was very complex in character, and for which I coined the term "polysynthetic." This contained about half a dozen different sulphides. The upper parts of the lode being partially oxidised, free sulphuric acid (H_2SO_4) was evolved. I therefore, following out a former discovery, added a little metallic zinc to the mercury in the boxes and on the plates with excellent results. When the free acid in the ore began to give out in the lower levels I added minute quantities of sulphuric acid to the water from time to time. I have since found, however, that with some water, particularly West Australian, the reaction is so feeble (probably owing to the lime and magnesia present) as to make this mode of treatment unsuitable.

HOW TO MAKE A DOLLY

I have seen some rather elaborate dollies, intended to be worked with amalgamating tables, but the usual prototype of the quartz mill is set up, more or less, as follows: A tree stump, from 9 in. to a foot diameter, is levelled off smoothly at about 2 ft. from the ground; on this is firmly fixed a circular plate of 1/2 in. iron, say 9 in. in diameter; a band of 3/16 in. iron, about 8 or 9 in. in height, fits more or less closely round the plate. This is the battery box. A beam of heavy wood, about 3 in. diameter and 6 ft. long, shod with iron, is vertically suspended, about 9 in. above the stump, from a flexible sapling with just sufficient spring in it to raise the pestle to the required height. About 2 ft. from the bottom the hanging beam is pierced with an augur hole and a rounded piece of wood, 1 1/2 in. by 18 in., is driven through to serve as a handle for the man who is to do the pounding. His mate breaks the stone to about 2 in. gauge and feeds the box, lifting the ring from time to time to sweep off the triturated gangue, which he screens through a sieve into a pan and washes off, either by means of a cradle or simply by panning. In dollying it generally pays to burn the stone, as so much labour in crushing is thus saved. A couple of small kilns to hold about a ton each dug out of a clay bank will be found to save fuel where firewood is scarce, and will more thoroughly burn the stone and dissipate the base metals, but it must be remembered that gold from burnt stone is liable to become so encrusted with the base metal oxides as to be difficult to amalgamate.

ROUGH WINDLASS

Make two St. Andrew's crosses with four saplings, the upper angle being shorter than the lower; fix these upright, one at each end of the shaft; stay them together by cross pieces till you have constructed something like a "horse," such as is used for sawing wood, the crutch being a little over 3 feet high. Select a leg for a windlass barrel, about 6 in. diameter and a foot longer than the distance between the supports, as straight as is procurable; cut in it two circular slots about an inch deep by 2 in. wide to fit into the forks; at one end cut a straight slot 2 in. deep across the face. Now get a crooked bough, as nearly the shape of a handle as nature has produced it, and trim it into right angular shape, fit one end into the barrel, and you have a windlass that will pull up many a ton of stuff.

PUDDLER

This is made by excavating a circular hole about 2 ft. 9 in. deep and, say 12 ft. in diameter. An outer and inner wall are then constructed of slabs 2 ft. 6 in. in height to ground level, the outer wall being thus 30 ft. and the inner 15 ft. in circumference. The circular space between is floored with smooth hardwood slabs or boards, and the whole made secure and water-tight. In the middle of the inner enclosure a stout post is planted, to stand a few inches above the wall, and the surrounding space is filled up with clay rammed tight. A strong iron pin is inserted in the centre of the post, on which is fitted a revolving beam, which hangs across the whole circumference of the machine and protrudes a couple of feet or so on each side. To this beam are attached, with short chains, a couple of drags made like V-shaped harrows by driving a piece of red iron through a heavy frame, shaped as a rectangular triangle.

To one end of the beam an old horse is attached, who, as he slowly walks round the circular track, causes the harrows and drags to so puddle the washdirt and water in the great wooden enclosure that the clay is gradually disintegrated, and flows off with the water which is from time to time admitted. The clean gravel is then run through a "cradle," "long Tom," or "sluice," and the gold saved. This, of course, is the simplest form of gold mining. In the great alluvial mines other and more intricate appliances are used but the principle of extraction is the same.

A MAKESHIFT PUMP

To make a temporary small "draw-lift" pump, which will work down to a hundred feet or more if required, take a large size common suction Douglas pump, and, after removing the top and handle, fix the pump as close to the highest level of the water in the shaft as can be arranged. Now make a square water-tight wooden column of slightly greater capacity than the suction pipe, fix this to the top of the pump, and by means of wooden rods, work the whole from the surface, using either a longer levered handle or, with a little ingenuity, horse-power. If you can get it the iron downpipe used to carry the water from the guttering of houses is more easily adapted for the pipe column; then, also, iron pump rods can be used but I have raised water between 60 and 70 feet with a large size Douglas pump provided only with a wooden column and rods.

SQUEEZING AMALGAM

For squeezing amalgam, strong calico, not too coarse, previously soaked in clean water, is quite as good as ordinary chamois leather. Some gold is fine enough to escape through either.

MERCURY EXTRACTOR

The mercury extractor or amalgam separator is a machine which is very simple in construction, and is stated to be most efficient in extracting quicksilver from amalgam, as it requires but from two to three minutes to extract the bulk of the mercury from one hundred pounds of amalgam, leaving the amalgam drier than when strained in the ordinary way by squeezing through chamois leather or calico. The principle is that of the De Laval cream separator—i.e., rapid centrifugal motion. The appliance is easily put together, and as easily taken apart. The cylinder is made of steel, and is run at a very high rate of speed.

The general construction of the appliance is as follows: The casing or receiver is a steel cylinder, which has a pivot at the bottom to receive the step for an upright hollow shaft, to which a second cylinder of smaller diameter is attached. The second cylinder is perforated, and a fine wire cloth is inserted. The mercury, after passing through the cloth, is discharged through the perforations. When the machine is revolved at great speed, the mercury is forced into the outside cylinder, leaving the amalgam, which has been first placed in a calico or canvas bag, in a much drier state than it could be strained by hand. While not prepared to endorse absolutely all that is claimed for this appliance, I consider that it has mechanical probability on its side, and that where large quantities of amalgam have to be treated it will be found useful and effective.

SLUICE PLATES

I am indebted to Mr. F. W. Drake for the following account of sluice plates, which I have never tried, but think the device worth attention:

"An addition has been made to the gold-saving appliances by the placing of what are called in America, 'sluice plates' below the ordinary table. The pulp now flows over an amalgamating surface, 14 ft. long by 4 ft. wide, sloping 1 1/2 in. to the foot, and is then contracted into a copper-plated sluice 15 ft. long by 14 in. wide, having a fall of 1 in. to the foot. Our mill manager (Mr. G. C. Knapp) advocated these sluice plates for a long time before I would consent to a trial. I contended that as we got little or no amalgam from the lower end of our table plates there was no gold going away capable of being recovered by copper plates; and even if it were, narrow sluice plates were a step in the wrong direction. If anything the amalgamating surface should be widened to give the particles of gold a better chance to settle. His argument was that the conditions should be changed; by narrowing the stream and giving it less fall, gold, which was incapable of amalgamation on the wide plates, would be saved. We finally put one in, and it proved so successful that we now have one at the end of each table. The per-centage recovered on the sluice plates, of the total yield, varies, and has been as follows:— October, 9.1 per cent; November, 6.9 per cent; December, 6.4 per cent; January, 4.3 per cent; February, 9.3 per cent."

MEASURING INACCESSIBLE DISTANCES

To ascertain the width of a difficult gorge, a deep river, or treacherous swamp without crossing and measuring, sight a conspicuous object at the edge of the bank on the farther side; then as nearly opposite and square as possible plant a stake about five feet high, walk along the nearer margin to what you guess to be half the distance across (exactitude in this respect is not material to the result), there plant another stake, and continuing in a straight line put in a third. The stakes must be equal distances apart and as nearly as possible at a right angle to the first line. Now, carrying in hand a fourth stake, strike a line inland at right angles to the base and as soon as sighting over the fourth stake, you can get the fourth and second stakes and the object on the opposite shore in line your problem is complete. The distance between No. 4 and No. 3 stakes is the same as that between No. 1 and the opposite bank.

TO SET OUT A RIGHT ANGLE WITH A TAPE

Measure 40 ft. on the line to which you wish to run at right angles, and put pegs at A and B; then, with the end of the tape held carefully at A, take 80 ft., and have the 80 ft. mark held at B. Take the 50 ft. mark and pull from A and B until the tape lies straight and even, you will then have the point C perpendicular to AB. Continue straight lines by sighting over two sticks in the well-known way.

Another method. — Stick a pin in each corner of a square board, and look diagonally across them, first in the direction of the line to which you wish to run at right angles, and then for the new line sight across the other two pins.

A SIMPLE LEVELLING INSTRUMENT

Fasten a common carpenter's square in a slit to the top of a stake by means of a screw, and then tie a plumb-line at the angle so that it may hang along the short arm, when the plumb-line hangs vertically and sights may be taken over it. A carpenter's spirit-level set on an adjustable stand will do as well. The other arm will then be a level.

Another very simple, but effective, device for finding a level line is by means of a triangle of wood made of half-inch boards from 9 to 12 ft. long. To make the legs level, set the triangle up on fairly level ground, suspend a plummet from the top and mark on the cross-piece where the line touches it. Then reverse the triangle, end for end, exactly, and mark the new line the plumb-line makes. Now make a new mark exactly half way between the two, and when the plumb-line coincides with this, the two legs are standing on level ground. For short water races this is a very handy method of laying out a level line.

TO MEASURE THE HEIGHT OF A STANDING TREE

Take a stake about your own height, and walking from the butt of the tree to what you judge to be the height of the timber portion you want, drive your stake into the ground till the top is level with your eyes; now lie straight out on your back, placing your feet against the stake, and sight a point on the tree. AB equals BC. If BC is, say 40 ft., that will be the height of your "stick of timber." Thus, much labour may be saved in felling trees the timber portion of which may afterwards be found to be too short for your purpose.

LEVELLING BY ANEROID BAROMETER

This should be used more for ascertaining relatively large differences in altitudes than for purposes where any great nicety is required. For hills under 2000 ft., the following rule will give a very close approximation, and is easily remembered, because 55 degrees, the assumed temperature, agrees with 55 degrees, the significant figures in the 55,000 factor, while the fractional correction contains *two fours*.

Observe the altitudes and also the temperatures on the Fahrenheit thermometer at top and bottom respectively, of the hill, and take the mean between them. Let B represent the mean altitude and b the mean temperature. Then 55000 X B - b/B + b = height of the hill in feet for the temperature of 55 degrees. Add 1/440 of this result for every degree the mean temperature exceeds 55 degrees; or subtract as much for every degree below 55 degrees.

TO DETERMINE HEIGHTS OF OBJECTS

By Shadows

Set up vertically a stick of known length, and measure the length of its shadow upon a horizontal or other plane; measure also the length of the shadow thrown by the object whose height is required. Then it will be: — As the length of the stick's shadow is to the length of the stick itself, so is the length of the shadow of the object to the object's height.

By Reflection

Place a vessel of water upon the ground and recede from it until you see the top of the object reflected from the surface of the water. Then it will be: — As your horizontal distance from the point of reflection is to the height of your eye above the reflecting surface, so is the horizontal distance of the foot of the object from the vessel to its altitude above the said surface.

Instrumentally

Read the vertical angle, and multiply its natural tangent by the distance between instrument and foot of object; the result is the height.

When much accuracy is not required vertical angles can be measured by means of a quadrant of simple construction. The arc AB is a quadrant, graduated in degrees from B to A; C, the point from which the plummet P is suspended, being the centre of the quadrant.

When the sights AC are directed towards any object, S, the degrees in the arc, BP, are the measure of the angle of elevation, SAD, of the object.

TO FIND THE DEPTH OF A SHAFT

Rule:—Square the number of seconds a stone takes to reach the bottom and multiply by 16.

Thus, if a stone takes 5 seconds to fall to the bottom of a shaft—

5 squared = 25; and 25 X 16 = 400 feet, the required depth of shaft.

DESCRIPTION OF PLAN FOR RE-USING WATER

Where water is scarce it may be necessary to use it repeatedly. In a case of this kind in Egypt, the Arab miners have adopted an ingenious method which may be adapted to almost any set of conditions. At a is a sump or water-pit; b is an inclined plane on which the mineral is washed and whence the water escapes into a tank c; d is a conduit for taking the water back to a; e is a conduit or lever pump for raising the water. A certain amount of filtration could easily be managed during the passage from c to a.

COOLING COMPOUND FOR HEATED BEARINGS

Mercurial ointment mixed with black cylinder oil and applied every quarter of an hour, or as often as expedient. The following is also recommended as a good cooling compound for heavy bearings:—Tallow 2 lb., plumbage 6 oz., sugar of lead 4 oz. Melt the tallow with gentle heat and add the other ingredients, stirring until cold.

CLEANING GREASY PLUMMER BLOCKS

When, through carelessness or unpreventable cause, plummer blocks and other detachable portions of machinery become clogged with sticky deposits of grease and impurities, a simple mode of cleansing the same is to take about 1000 parts by weight of boiling water, to which add about 10 or 15 parts of ordinary washing soda. Keep the water on the boil and place therein the portions of the machine that are to be cleaned; this treatment has the effect of quickly loosening all grease, oil, and dirt, after which the metal is thoroughly washed and dried. The action of the lye is to form with the grease a soap soluble in water. To prevent lubricating oil hardening upon the parts of the machinery when in use, add a third part of kerosene.

AN EXCELLENT ANTI-FRICTION COMPOUND

For use on cams and stamper shanks, which will be harmless should it drop into the mortar or stamper boxes, is graphite (black-lead) and soft soap. When the guides are wooden, the soft soap need not be added; black-lead made into a paste with water will act admirably.

TO CLEAN BRASS

Oxalic acid 1 oz., rotten stone 6 oz., powdered gum arabic 1/2 oz., sweet oil 1 oz. Rub on with a piece of rag.

A SOLVENT FOR RUST

It is often very difficult, and sometimes impossible, to remove rust from articles made of iron. Those which are very thickly coated are most easily cleaned by being immersed in a nearly saturated solution of chloride of tin. The length of time they remain in this bath is determined by the thickness of the coating of rust. Generally from twelve to twenty-four hours is long enough.

TO PROTECT IRON AND STEEL FROM RUST

The following method is but little known, although it deserves preference over many others. Add 7 oz. of quicklime to 1 3/4 pints of cold water. Let the mixture stand until the supernatant fluid is entirely clear. Then pour this off, and mix with it enough olive oil to form a thick cream, or rather to the consistency of melted and re-congealed butter. Grease the articles of iron or steel with this compound, and then wrap them up in paper, or if this cannot be done, apply the mixture somewhat more thickly.

TO KEEP MACHINERY FROM RUSTING

Take 1 oz. of camphor, dissolve it in 1 lb. of melted lard; mix with it (after removing the scum) as much fine black-lead as will give it an iron colour; clean the machinery, and smear it with this mixture. After twenty-four hours rub off and clean with soft, linen cloth. This mixture will keep machinery clean for months under ordinary circumstances.

FIRE-LUTE

An excellent fire-lute is made of eight parts sharp sand, two parts good clay, and one part horse-dung; mix and temper like mortar.

ROPE-SPLICING

A short splice is made by unlaying the ends of two pieces of rope to a sufficient length, then interlaying them, draw them close and push the strands of one under the strands of the other several times. This splice makes a thick lump on the rope and is only used for slings, block-straps, cables, etc.

www.ingramcontent.com/pod-product-compliance
Lightning Source LLC
Chambersburg PA
CBHW031417210526
45464CB00005B/1933